ICTAM

ICTAM

Edited by

P. Lugner and K. Hedrick

Selected papers from the 20th International Congres of Theoretical and Applied Mechanics held in Chicago, 28 August-1 September 2000

Supplement to Vehicle Systems Dynamics, Volume 35

SWETS & ZEITLINGER PUBLISHERS

LISSE ABINGDON EXTON (PA) TOKYO

Library of Congress Cataloging-in- Publication Data

(applied for)

Printed in the Netherlands by Wilco bv, Amersfoort

ISBN 90 265 1812 9

TABLE OF CONTENTS

Vehicle System Dynamics Supplement 35 (2001), pp. 1

Introduction

The Congress Committee of the International Union of Theoretical and Applied Mechanics (IUTAM) selected a proposal from the International Association of Vehicle System Dynamics (IAVSD), one of its Affiliated Organizations, on the topic "Vehicle System Dynamics" for one of the six Minisymposia to be included in the 20th International Congress of Theoretical and Applied Mechanics (ICTAM 2000), Chicago August 28th to September 1st 2000. One of the reasons for this selection was the increasing interest of vehicle engineers in the application of computational methods developed in theoretical mechanics. Furthermore, Chicago is close to the centers of the North American automobile industry providing an excellent opportunity to discuss joint problems from an engineering and scientific point of view. As a matter of fact, the Minisymposium on vehicle dynamics was very well attended.

The Minisymposium comprised introductory lectures and one sectional lecture in the main sequence of the conference. Since these lectures by invited scientists were printed in the proceedings of the conference in full length, only a short survey is given in this supplement. All other congress papers had to pass the rigorous conference selection procedure to be accepted for presentation and were presented in printed form by short abstracts.

To provide information on the other Contributed Papers in this Minisymposium and conference papers with a strong connection to vehicle system dynamics, a restricted number was selected after further reviewing and printed in this present supplement of the VSD-Journal. The authors therefore revised their lecture presentations and extended them with additional wording and figures.

We hope that this selection of scientific contributions illustrate the progress in vehicle dynamics at the turn of the century.

The editors: P. Lugner, K. Hedrick

Vehicle System Dynamics Supplement 35 (2001), pp. 3-5

Survey of the Sectional and Introductory Lectures

WERNER SCHIEHLEN *

The scientific program of ICTAM 2000 related to the minisymposium "Vehicle System Dynamics" comprised one sectional lecture by T.J. Gordon (UK), four introductory lectures presented by W. Schiehlen (Germany), R. Anderson (Canada), R. Sharp (UK) and M. Abe (Japan), and quite a number of contributed papers presented in lecture and seminar sessions. The sectional and introductory lectures are published in the Congress proceedings by Kluwer Academic Publishers and they are summarized here.

The sectional lecture by T.J. Gordon, Department of Aeronautical and Automotive Engineering, Loughborough University, U.K. was entitled "Adaptive, Nonlinear and Learning Techniques for the Control of Vehicle Ride Dynamics". This paper is concerned with the ride dynamics and the control of whole-body vibration of road vehicles, to provide comfort and vibration isolation for occupants and transported goods. Ride isolation from road unevenness is conventionally achieved through the pneumatic tire, coupled with a spring and damper in the suspension; however substantial benefits can be derived from computer control of the suspension system. Active ride control also benefits from on-line adaptation, and similar advantage can be derived via nonlinear feedback control. This paper reviews the fundamental issues and considers the potential for future vehicles, against a background of increasing total system complexity and interaction, as well as the continuing need for robust, safe and fault-tolerant operation. Consideration is also given to the use of intelligent control systems that adapt and learn in real-time on the vehicle.

The first introductory lecture on "Software Tools: From Multibody System Analysis to Vehicle System Dynamics" was based on a paper coauthored by W. Kortüm, Institute of Aeroelasticity, DLR German Aerospace Center, Wessling, Germany, W. Schiehlen, Institute B of Mechanics, University of Stuttgart, Stuttgart, Germany and M. Arnold, Institute of Aeroelasticity, DLR German Aerospace Center, Wessling, Germany.

This paper reports on the transition of theoretical research in mechanics to the everyday application of the developed tools in industry. After successful application to spacecrafts and appropriate theoretical examinations, multibody

*University of Stuttgart, Pfaffenwaldring 9, 70550 Stuttgart, Germany

system approaches, their formalisms and software became of interest to the vehicle system dynamicists both for rail and road vehicles. This introductory paper sketches a few important milestones in the general development of multibody systems, related to vehicle system dynamics. While at the beginning the absence of vehicle specific force laws was the major stumbling block, later on the numerical methods and the efficiency of the formalisms became of prime focus. In the recent past, the transition from system analysis to system design and optimization as well as the integration into multidisciplinary computer aided engineering of vehicle systems was and still is a challenge.

The second introductory lecture on "Rail Vehicle Dynamics for the 21st Century" was based on a paper coauthored by R.J. Anderson, Queen's University, Kingston, Ontario, Canada, J.A. Elkins, RVD Consulting Inc., Pueblo, CO, USA and B.V. Brickle, Transportation Technology Center Inc., Pueblo, CO, USA.

Although railway vehicles were first used in the 18th century, it was only in the 20th century that engineers began to understand their dynamics and were able to write down and solve their equations of motion. The single most fundamental property of a railway vehicle is the steel wheel running on a steel rail with all the traction, braking and guidance forces being transmitted through a small contact area between the wheel and the rail. While the inherent guidance provided by a conical wheelset running on rails was known even to George Stephenson in 1821, it is only in the last 40 years that an adequate theory has been available to vehicle suspension designers. It is well known that a mechanical system which is subject to nonconservative forces may become dynamically unstable under certain conditions. In a railway vehicle the nonconservative forces arise at the contact point between the wheel and rail due to creepage, and the sustained lateral and yawing oscillation is known as hunting. This paper traces the developments in railway dynamics up to the end of the 20th century and discusses the latest developments, with predictions as to what the future might hold.

The third introductory lecture presented by R.S. Sharp, School of Mechanical Engineering, Cranfield University, UK, was entitled "Fundamentals of the Lateral Dynamics of Road Vehicles".

In this paper road vehicle steering control is discussed as a man-machine problem. Rolling contact between tire and ground is central to understanding of the engineering of the vehicle and a section is devoted to an explanation of it. The steady turning behaviour of road vehicles is extremely significant. The interactions of suspensions and chassis have a strong influence on the system qualities and these are highlighted. The full dynamics are then described by reference firstly to the well-known yaw/sideslip linear model and, then, by generalization of the results to include continuous nonlinear behaviour of the simple car and nonlinear behaviour of more complex, higher order systems representing vehicles with much design detail. Linearization about a trim, equilibrium cornering condition and the intimate relationship between steady turning behaviour and static stability properties are presented. Vehicle design alternatives are exposed and the manner in which electronic chassis control systems and the use of the tire longitudinal force capability to avoid necessary compromise solutions in the passive world are mentioned. The paper concludes with a brief consideration of trucks, truck-trailer trains and single track vehicles.

The fourth introductory lecture on "A Mechatronic Approach to Advanced Vehicle Control Design" was based on a paper coauthored by Masato Abe, Kanagawa Institute of Technology, Japan and Karl Hedrick, University of California, Berkeley, USA.

The first part of this paper focuses on vehicle chassis control, primarily to improve vehicle handling performance and safety. One way to improve handling performance through chassis control uses four wheel steering. This approach depends upon the tire lateral forces, which are proportional to the steer angle in a range where the lateral acceleration is small. However, in a higher lateral acceleration range the lateral force becomes strongly dependent upon the tire vertical load and longitudinal force as well, so that the control law becomes rather sensitive to the vehicle motion and it's environmental conditions. A direct yaw moment control controlling vehicle motion by a yaw moment is another approach. One of the major advantages of this control method is that the tire longitudinal force has small feedback from the vehicle lateral motion as long as it is within the limit of the tire capacity due to the vertical load. Therefore, a precise yaw moment, required to control the vehicle lateral motion, can be developed and the vehicle motion with the control becomes more robust to fluctuations in the vehicle running condition and it's environment. The second part of this paper discusses the use of preview information to improve the ride quality of off-road vehicles. This part presents an interesting control methodology called Model Predictive Control that uses real-time, on-line constrained optimization to improve ride quality subject to constraints such as available suspension clearances.

Vehicle System Dynamics Supplement 35 (2001), pp. 7-22

Influence of Locomotive Tractive Effort on the Forces Between Wheel and Rail

OLDRICH POLACH[1]

SUMMARY

For the complex simulation of dynamic behaviour of a locomotive in connection with drive dynamics and traction control, large longitudinal creep between wheel and rail and decreasing part of creep-force function behind adhesion limit have to be modelled. This needs an adaptation of models used for the computation of tangential forces between wheel and rail in vehicle dynamics. A fast method for the calculation of wheel-rail forces developed by the author was extended to get the properties mentioned above. The simulations with a locomotive model running in a curve with varying tractive force show the influence on the wheel-rail forces distribution and on the radial steering ability. Another application example presents the investigation of a complex mechatronic system of the traction control and vehicle dynamics. A comparison with the measurements confirms that the proposed method is suitable for the investigation of problems regarding the interaction of axle drive dynamics and vehicle behaviour.

1. INTRODUCTION

The standard methods of vehicle dynamics investigate only a part of the complex mechatronic systems of railway traction vehicles. To investigate e. g. influence of locomotive tractive effort on the forces between wheel and rail in the curve, influence of traction control parameters on the wheel-rail forces, on the dynamic behaviour of the locomotive or of the train set etc., the mechanics, electric and electronics have to be simulated simultaneously. The model of wheel-rail forces used in such computer simulations should contain the properties necessary for vehicle dynamics as well as for axle drive and traction control dynamics. This is not possible using the standard methods and needs an adaptation of models used for the computation of tangential forces between wheel and rail in vehicle dynamics. A fast method for the calculation of wheel-rail forces developed by the author was extended to get the properties mentioned above and verified by comparisons with measurements.

Application examples illustrate the influence of locomotive tractive effort on the forces between wheel and rail. Another application example presents the investigation of a complex mechatronic system of the traction control and vehicle dynamics using the co-simulation of ADAMS/Rail and MATLAB-SIMULINK.

[1] Adtranz Switzerland, Zürcherstr. 41, CH-8401 Winterthur, Switzerland

2. MODELS OF WHEEL-RAIL FORCES IN THE VEHICLE DYNAMICS SIMULATIONS

There are various methods for the computation of wheel-rail forces in railway vehicle dynamics. The best known methods can be divided into four groups:

- exact theory by Kalker (programme CONTACT)
- simplified theory by Kalker (programme FASTSIM)
- look-up tables
- simplified formulae and saturation functions.

The exact theory by Kalker (computer programme CONTACT [5]) has not been used in the simulations because of its very long calculation time.

The simplified theory used in Kalker's programme FASTSIM [3] is much faster than the exact theory, but the calculation time is still relatively long for on-line computation in complicated multi-body systems. FASTSIM is used e. g. in the railway vehicle simulation tools ADAMS/Rail, MEDYNA, SIMPACK, GENSYS, VOCO.

Another possibility for computer simulations consists in the use of look-up tables with saved pre-calculated values (ADAMS/Rail, VAMPIRE). Because of the limited data in the look-up table, there are differences to the exact theory as well. Large tables are more exact, but the searching in such large tables consumes calculation time.

Searching for faster methods some authors found approximations based on saturation functions (e.g. Vermeulen-Johnson [12], Shen-Hedrick-Elkins [11]). The calculation time using these approximations is short, but there are significant differences to the exact theory especially in the presence of spin. Simple approximations are often used as a fast and less exact alternative to standard methods (e. g. in MEDYNA, VAMPIRE, SIMPACK).

A very good compromise between the calculation time and the required accuracy allows a fast method for the computation of wheel-rail forces developed by the author [7], [8]. In spite of the simplifications used, spin is taken into consideration. Due to the short calculation time, the method can be used as a substitute for Kalker's programme FASTSIM to save computation time or instead of approximation functions to improve the accuracy.

The proposed method has been used in the simulations in different programmes since 1990. The algorithm is implemented in ADAMS/Rail as an alternative parallel to FASTSIM and to the Table-book. The calculation time is faster than FASTSIM's and usually faster than the Table-book as well. The proposed algorithm provides a smoothing of the contact forces in comparison with the Table-book and there are no convergence problems during the integration. The programme was tested as user routine in programmes: SIMPACK, MEDYNA, SIMFA and various user's own programmes with very good experience.

The proposed method assumes an ellipsoidal contact area with half-axes a, b and normal stress distribution according to Hertz. The maximum value of tangential stress τ at any arbitrary point is

$$\tau_{\max} = f \cdot \sigma \tag{1}$$

where f - coefficient of friction,
σ - normal stress.

The coefficient of friction f is assumed constant in the whole contact area. The solution described in [7], [8] obtains the resultant tangential force (without spin) as

$$F = -\frac{2 \cdot Q \cdot f}{\pi}\left(\frac{\varepsilon}{1+\varepsilon^2} + \arctan \varepsilon\right) \tag{2}$$

where Q - wheel load
f - friction coefficient
ε - gradient of the tangential stress in the area of adhesion

$$\varepsilon = \frac{2}{3}\frac{C \cdot \pi \cdot a^2 \cdot b}{Q \cdot f} s \tag{3}$$

where C - proportionality coefficient characterising the contact elasticity
a, b - half-axes of the contact ellipse
s - creep

$$s = \sqrt{s_x^2 + s_y^2} \tag{4a}$$

$$s_i = \frac{w_i}{V} \qquad i = x, y \tag{4b}$$

where s_x, s_y - creep in longitudinal (x) and lateral (y) directions
w_x, w_y - creep (slip) velocity in longitudinal (x) and lateral (y) directions
V - vehicle speed.

Using the Kalker's linear theory to express the coefficient C, the equation (3) has then the form (in the case of only longitudinal creep)

$$\varepsilon = \frac{1}{4}\frac{G \cdot \pi \cdot a \cdot b \cdot c_{11}}{Q \cdot f} s_x \tag{5}$$

where c_{11} - Kalker's coefficient for longitudinal direction [2].

The calculation of tangential forces for general case with combination of longitudinal and lateral creep and spin allows the computer code published in [8].

The proposed method was verified by making a comparison between curving behaviour calculations using for the computation of wheel-rail forces the programme FASTSIM and the proposed method, and by comparison with measurements. A model of the four axle SBB 460 locomotive of the Swiss Federal Railways was built by means of the simulation tool ADAMS/Rail. The locomotive design combines very good curving performance with high maximal speed due to the coupling of wheelsets, realised by a mechanism with a torsional shaft assembled to the bogie frame. The model used in simulations consists of 51 rigid bodies and contains 266 degrees of freedom. The results using both calculation methods

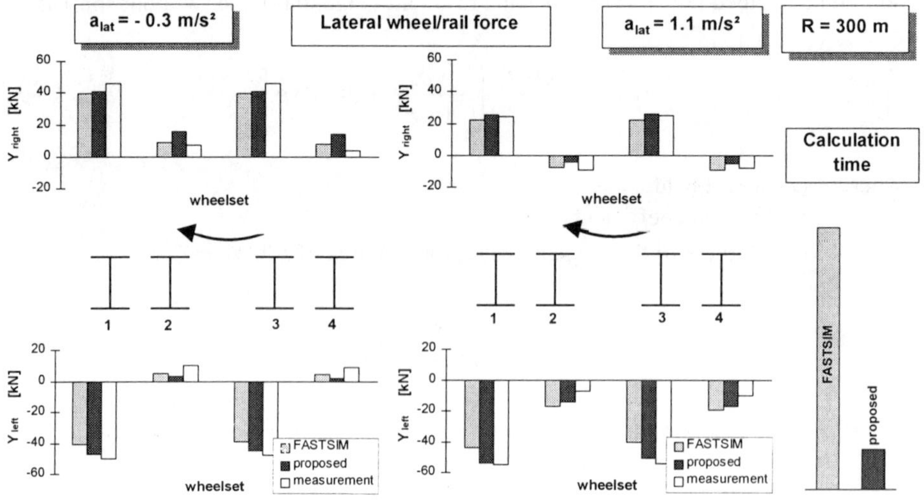

Fig. 1. Measured lateral wheel-rail forces in a curve with 300 m radius compared with the simulations using the method developed by the author [8] and using FASTSIM.

mentioned are similar, see Fig. 1. However, there is a significant difference in the calculation time. The results computed using the proposed method show good agreement with the measurements. Especially in the case of the leading wheelsets, they are nearer to the actually measured values than the results obtained in the simulations using FASTSIM.

3. DIFFERENCES BETWEEN THE VEHICLE DYNAMICS AND AXLE DRIVE DYNAMICS

Depending on the aim of the tests, different measured creep-force functions can be found in the literature [6]. Because of the variety of measurements different models are used for the same physical phenomenon - forces between wheel and rail - in the vehicle dynamics and axle drive dynamics (Fig. 2). How is this possible? The reason for this are different fields of parameters and different areas of investigations.

In the vehicle dynamics small creep values are of main importance. Tangential forces in longitudinal as well as in lateral directions influence the vehicle behaviour, therefore longitudinal and lateral creep as well as spin should be taken into account. Based on the theory of rolling contact, the creep forces depend on the creep as non-dimensional value. The friction coefficient is assumed to be constant. The difference between dry and wet conditions is usually expressed only with the value of friction coefficient and not with a change of the creep-force function gradient.

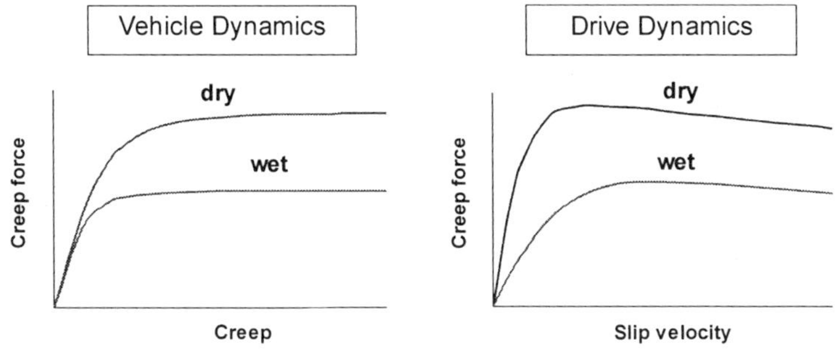

Fig. 2. Differences between the typical creep-force functions used in the vehicle dynamics and in axle drive dynamics.

In the drive dynamics large values of longitudinal creep influence its behaviour. In the simulated systems usually only the longitudinal direction is taken into account. Based on the experiments, the creep forces are usually assumed as dependent on the slip velocity between wheel and rail. There is a maximum of creep-force function, so called adhesion optimum, and a decreasing section behind this maximum. The gradient and the form of creep-force functions for wet, dry or other conditions are different.

4. MODEL OF WHEEL-RAIL FORCES USEFUL FOR SIMULATION OF VEHICLE DYNAMICS AND AXLE DRIVE DYNAMICS INTERACTION

For the complex simulation of dynamic behaviour of locomotive or traction vehicle in connection with drive dynamics and traction control, high longitudinal creep and decreasing section of creep-force function behind the adhesion limit has to be taken into account. Thc different wheel-rail models described above have to be made into one model, which is suitable for both vehicle dynamics and drive dynamics simulations.

A creep-force law with a marked adhesion optimum can be modelled using the friction coefficient decreasing with increasing slip (creep) velocity between wheel and rail. The dependence of friction on the slip velocity was observed by various authors and is described e. g. in [6], [9]. The variable friction coefficient can be expressed by the following equation

$$f = f_0 \cdot \left[(1 - A) \cdot e^{-B \cdot w} + A\right] \qquad (6)$$

where f_0 - maximum friction coefficient

w - magnitude of the slip (creep) velocity vector [m/s]
B - coefficient of exponential friction decrease [s/m]
A - ratio of limit friction coefficient f_∞ at infinity slip velocity to maximum friction coefficient f_0
$A = f_\infty / f_0$.

From the point of view of tractive effort and traction dynamics the creep law for bad adhesion conditions is of main importance: wet rail and surface pollution e. g. oil, dirt, moisture. Even for dry conditions the gradient of creep-force functions is usually lower than the theoretical value. The reason of this is the layer of moisture, which can be taken into consideration in the "stiffness coefficient" of surface soil [4]. In the vehicle dynamics, this real conditions are taken into account simply with a reduction of Kalker's creep coefficients [2]. This method is used for the linear creep force law but can be used generally as well.

The combination of
- the friction coefficient as a function of slip velocity between wheel and rail and
- the reduction of Kalker's creep coefficients

allows
- on the one hand to achieve the form of creep-force function with an adhesion optimum as known from measurements and as necessary for traction investigations
- and on the other hand to keep the principles of creep forces computation in dependence on longitudinal and lateral creep and spin.

In this manner, the creep-force functions can be adapted for various conditions of wheel-rail contact, see Fig. 3. The diagrams show the adhesion coefficient $\mu_x = F_x/Q$ in function of longitudinal creep s_x.

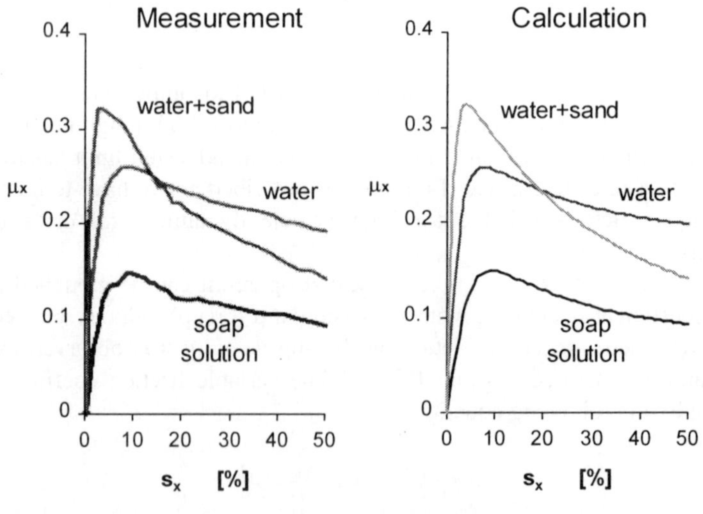

Fig. 3. Modelling of various adhesion conditions using the friction coefficient as function of slip velocity between wheel and rail (measurement: adhesion tests with the SBB 460 locomotive, $V = 20$ km/h).

At the same time, using this principle, the adhesion maximum decreases and parallel to this the creep value at this maximum decreases with increasing of vehicle speed *V*, which is a phenomenon well known from measurements. It can be observed on Fig. 4, which shows example of creep-force functions modelled using the method according to Polach [8] with the extension described above. The same

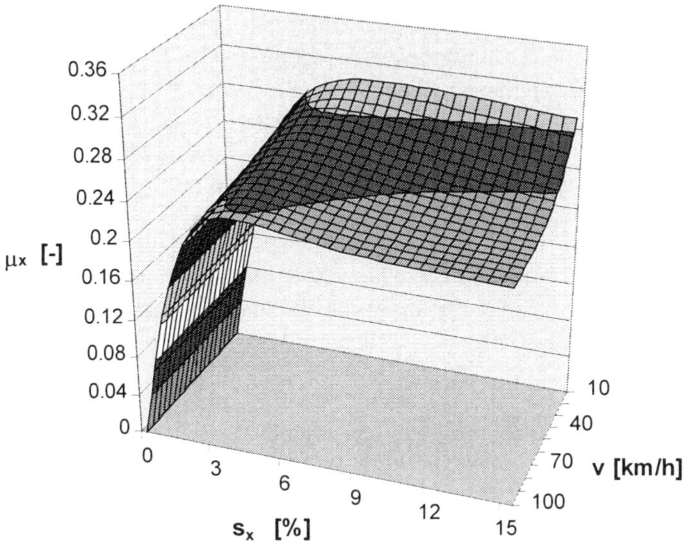

Fig. 4. Influence of vehicle speed on the form of creep-force functions using the friction coefficient as function of slip velocity between wheel and rail.

principles can be used to extend the model of wheel-rail forces based on FASTSIM or any other method of creep force computation.

Although the method described above allows to reproduce the basic tendencies of measured creep-force functions at large creep values, there are some limitations. To achieve the adhesion optimum at large creep values, a significant reduction of Kalker's coefficients would be necessary (less than 0.1). But the measurements of the creep-force function gradient do not usually show so low values.

The explanation may lie in the combination of dry and wet friction. For small creep values, the area of adhesion extends to the greater part of the contact area. The conditions are similar to dry friction. For large creep values, there is slip in the main part of the contact area. The layer of water or pollution influences the resultant force. The "stiffness coefficient" of surface soil decreases and, as a result of this, the creep-force function changes its gradient significantly.

In spite of the complexity of this phenomenon, a good possibility was found to model the wheel-rail contact forces in good agreement with the measured functions using a simple extension of the method developed by the author [8]. In this method the tangential forces (equation (2)) are expressed as a function of the gradient ε of the tangential stress (equation (5)). Using a reduction of Kalker's coefficient with a factor k, the equation (5) has the form

$$\varepsilon_R = \frac{1}{4}\,\frac{G\cdot\pi\cdot a\cdot b\cdot k\cdot c_{11}}{Q\cdot f}\,s_x = k\cdot\varepsilon \tag{7}$$

In (2) there are two terms: one of them connected to the area of adhesion, another one to the area of slip. Using different reduction factors k_A in the area of adhesion and k_S in the area of slip, the equation (2) has the form

$$F = -\frac{2\cdot Q\cdot f}{\pi}\left(\frac{k_A\cdot\varepsilon}{1+(k_A\cdot\varepsilon)^2}+\arctan(k_S\cdot\varepsilon)\right) \tag{8}$$

Using $k_A > k_S$, there is nearby no reduction of the creep-function gradient at small creepages but a significant reduction of the gradient near saturation. The gradient of the creep-force function at the co-ordinate origin corresponds to the reduction of Kalker's coefficient

$$k = \frac{k_A + k_S}{2} \tag{9}$$

The creep-force functions are shown in Fig. 5 for a constant coefficient of friction. Using the proposed method with two reduction factors, the form of the creep-force functions is more similar to the form of measured functions than using one reduction factor k, see Fig. 5. The diagrams show creep-force functions for longitudinal direction in non-dimensional co-ordinates

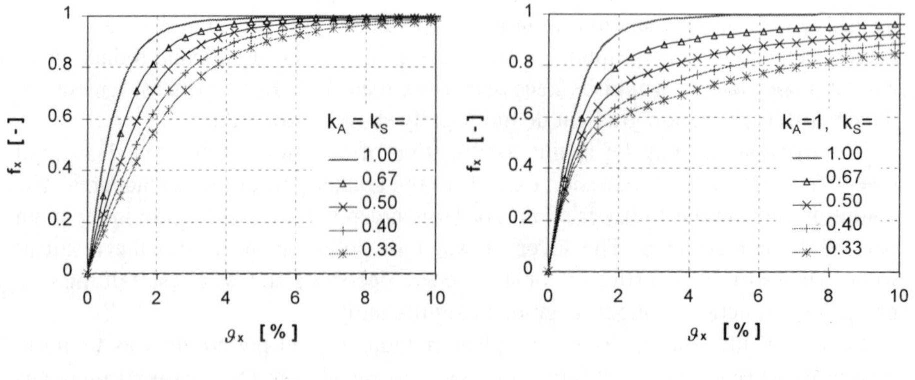

Fig. 5. Reduction of Kalker's coefficients: the standard method (left) and the proposed method with two different reduction factors (right).

- non-dimensional wheel-rail force

$$f_x = \frac{F_x}{Q \cdot f} \tag{10}$$

- non-dimensional longitudinal creep

$$\vartheta_x = \frac{G.a.b.c_{11}.s_x}{Q.f} \tag{11}$$

Using the proposed method various measured creep-force functions were modelled. Fig. 6 shows measurements with the Adtranz SBB 460 locomotive. There

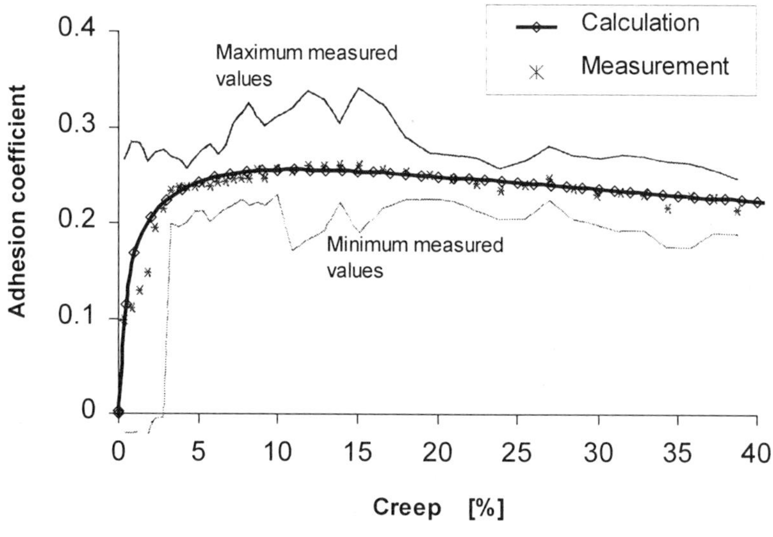

Fig. 6. Model of wheel-rail contact forces based on the measurements with the Adtranz SBB 460 locomotive (7 measurements on wet rails; $V = 40$ km/h; model parameters: $k_A = 0.16$, $k_S = 0.07$, $f_0 = 0.305$, $A = 0.50$, $B = 0.16$ s/m).

are results of 7 measurements with speed of 40 km/h on wet rail. The measured points were classified and average values, maximal and minimal values were calculated. Using the model described, parameters were found which provide a creep-force model with a very good agreement with the measurements.

Another example in Fig. 7 shows measurements with the Siemens Eurosprinter locomotive No. 127 001 (according to [1]). Changing the parameters a very good agreement of the measured and modelled dependence can be achieved again.

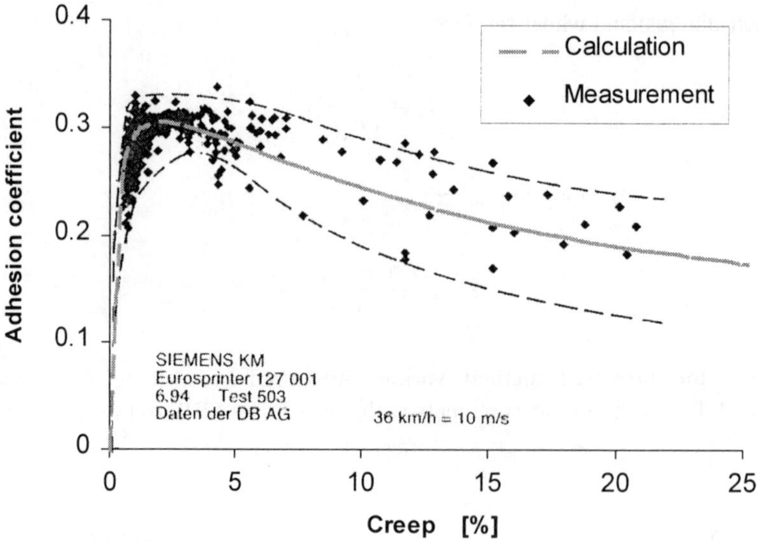

Fig. 7. Model of wheel-rail contact forces based on the measurements with the Siemens locomotive Eurosprinter 127 001 [1] (rail conditions not mentioned; V = 36 km/h; model parameters: $k_A = 0.72$, $k_S = 0.36$, $f_0 = 0.36$, $A = 0.38$, $B = 0.70$ s/m).

The wheel-rail models with parameters presented above were tested for various speeds and for range of longitudinal creep from very small to high creep values. Even in this large range the results are plausible and the model is not limited to one speed or to a small creep range. Of course, change of conditions in contact of wheel and rail as well as other effects, e. g. the cleaning effect due to large creep (so called rail conditioning) will cause change of wheel-rail model parameters.

The examples confirm that the method gives a very good possibility to model the functionality of wheel-rail contact necessary for the investigation of axle drive systems. Parallel to this the method allows to keep the principles of creep forces computation in dependence on longitudinal and lateral creep and spin, which is necessary for vehicle dynamics investigations.

5. APPLICATION EXAMPLES WITH THE PROPOSED WHEEL-RAIL MODEL

As an application example of the proposed wheel-rail model, two cases will be presented

- Quasi-static and dynamic investigations of the influence of locomotive traction effort on the wheel-rail forces and curving behaviour
- Influence of traction control on locomotive dynamic behaviour using co-simulation of ADAMS/Rail and MATLAB/SIMULINK.

5.1. Influence of Locomotive Tractive Effort on the Curving Behaviour

In order to verify the method of wheel-rail forces computation, the ADAMS/Rail model of the SBB 460 locomotive was extended by a full model of the driving system and runs through a curve on wet rails were simulated. Fig. 8 shows the

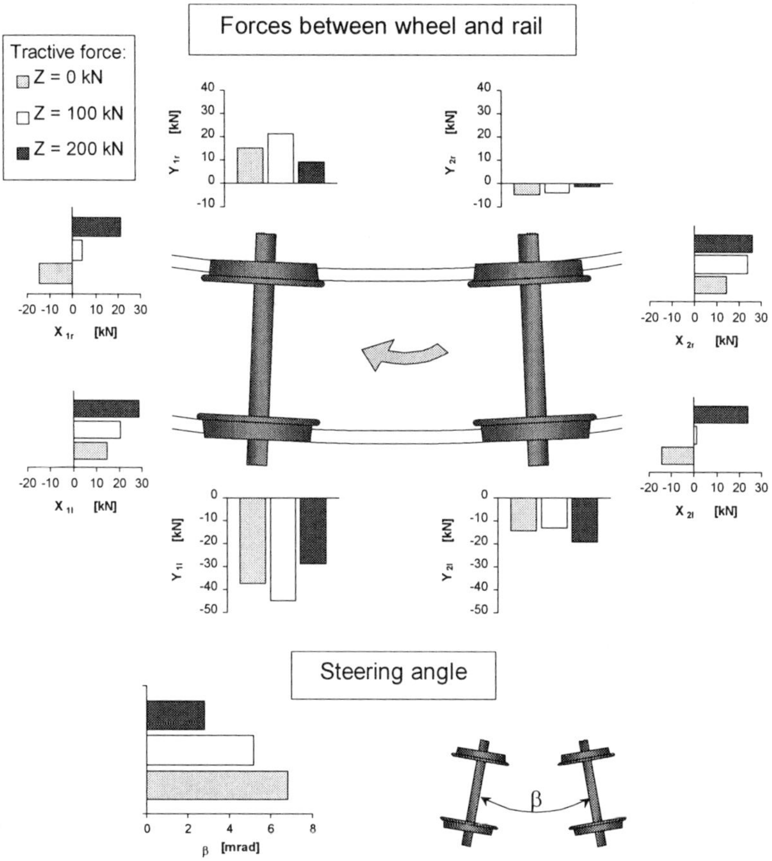

Fig. 8. Longitudinal and lateral wheel-rail forces and steering angle between wheelsets for different tractive effort.

distribution of longitudinal and lateral quasi-static wheel-rail forces in a curve of 300 m radius, running at a speed of 87.7 km/h (lateral acceleration $a_{lat} = 1.0$ m/s²). The wheel-rail forces change with increasing tractive effort. Without tractive effort, there are opposite longitudinal forces on the left and right wheel. With increasing tractive effort the longitudinal forces on the outer wheel of the first wheelset and on the inner wheel of the second wheelset increase. An even distribution of the longitudinal forces will be achieved first on the adhesion limit. The angle between the wheelsets in the horizontal plane (so called steering angle) depends on the tractive effort as well.

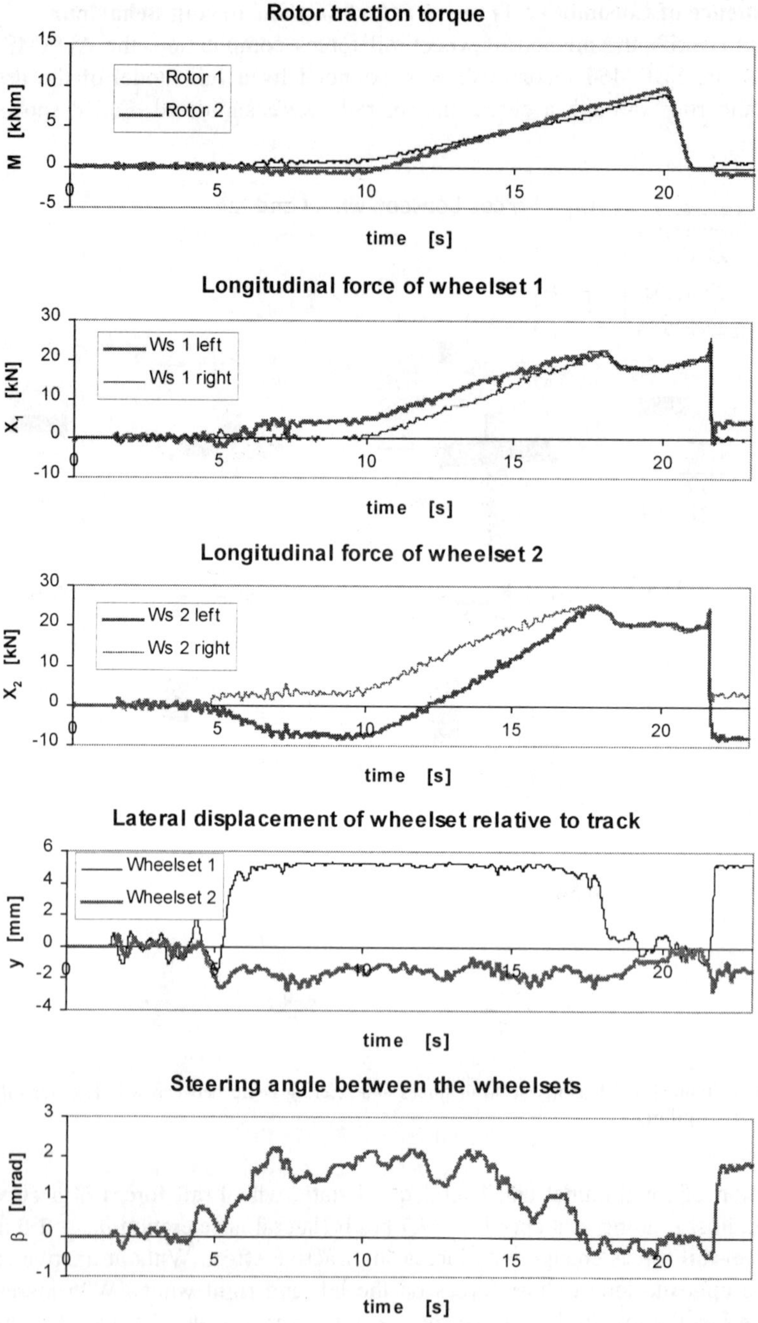

Fig. 9. Time plots of adhesion test simulation (curve radius $R = 400$ m, speed $V = 70$ km/h).

Without tractive effort, the angle between the wheelsets achieves the maximal value due to the steering moment of longitudinal forces. With increasing tractive effort the distribution of longitudinal forces changes as explained before, the steering moment acting on the wheelsets decreases and the steering angle between the wheelsets decreases as well. In spite of this the experience with these locomotives confirms that there is no negative influence on the wear of wheels and rails because of the statistically rare occurrence of maximal tractive effort. On the Gotthard-route the locomotives achieve 3 to 4 times longer intervals between the renewals of wheel profiles than the previous ones [13].

In order to test the possibility of simulating the dynamic change of traction torque, a test case used during the adhesion test was simulated. In the course of curving simulation, the traction torque was increased from zero to the adhesion limit, in a similar way as during the adhesion test measurements. In this manner, a run on the unstable (decreasing) section of creep-force function was simulated. The influence of increased tractive effort on longitudinal and lateral wheel-rail forces and the steering angle between the wheelsets of a single bogie were investigated. The simulation results show the same behaviour as observed during the tests. The time plots of traction torque, longitudinal wheel-rail forces, lateral displacements of the wheelsets relative to track and steering angle between wheelsets for the speed of 70 km/h are shown in Fig. 9. With increasing tractive effort, the wheelset steering

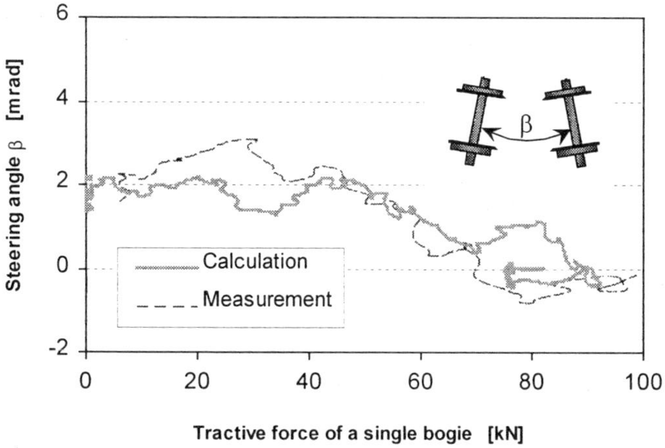

Fig. 10. Comparison of the measured and calculated steering angle between wheelsets as function of tractive effort of a single bogie.

ability decreases. Simultaneously, the first wheelset of the bogie moves to the inner rail. The calculation shows the same dynamic behaviour tendency due to the change of the tractive effort as observed during the tests. The calculated steering angle as a function of tractive force in the steering linkages is compared with the measurement in Fig. 10. The comparison confirms a good agreement in the tendency although the simulated track irregularity does not correspond to the reality.

5.2. Influence of Traction Control on Locomotive Dynamic Behaviour

The traction control influences possible tractive effort as well as the dynamic behaviour of the locomotive. Because of different simulation tools used in traction control and in vehicle dynamics, a co-simulation was chosen as a way to investigate this mechatronic system. The control part of the locomotive modelled in MATLAB/SIMULINK comprises the motor control and the adhesion control. The motor control is represented by measured transfer functions and the adhesion control is represented by SIMULINK blocks.

The adhesion controller uses an adaptive control system [10] which determines the optimal longitudinal slip. The control is based on the rotor velocity response to the torsion oscillations, excited by a test signal superimposed to the motor torque. Based on the filtered sinusoidal rotor velocity, the gradient of the creep-force function is computed. The adhesion optimum is achieved when the gradient of creep-force function is zero.

The ADAMS/Rail model of the SBB 460 locomotive described above was connected with the control unit in MATLAB/SIMULINK and the system was investigated using co-simulation. The input values of the ADAMS/Rail model consisted of the torque of the motors, and the input values of the control unit in MATLAB/SIMULINK were the angular velocities of the rotors and the desired motor torque value.

As an example of computed results a simulation of locomotive starting and acceleration is presented. The track consists of a short straight, a right curve with $R = 300$ m, another short straight and another right curve with $R = 385$ m. Fig. 11 shows the following computed values in function of the distance along the track

- Locomotive speed
- Torque of rotor 1 and 2
- Longitudinal forces of the first wheelset
- Lateral forces of the first wheelset
- Steering angles between the wheelsets of the first and second bogie.

The rotor torque oscillations due to the test signal can be seen in the results. Because only the first bogie provided the tractive effort, a difference in the steering angle of the first and second bogie due to traction can be observed. At the same time there are changes in the longitudinal and lateral wheel-rail forces due to track curvature and due to increasing vehicle speed.

The results of presented application examples confirm this as a possible way for the simulation of complex problems of vehicle dynamics, axle drive dynamics and traction control. The next step will be an optimisation of locomotive behaviour using a model, which contains mechanics, electrics and controls.

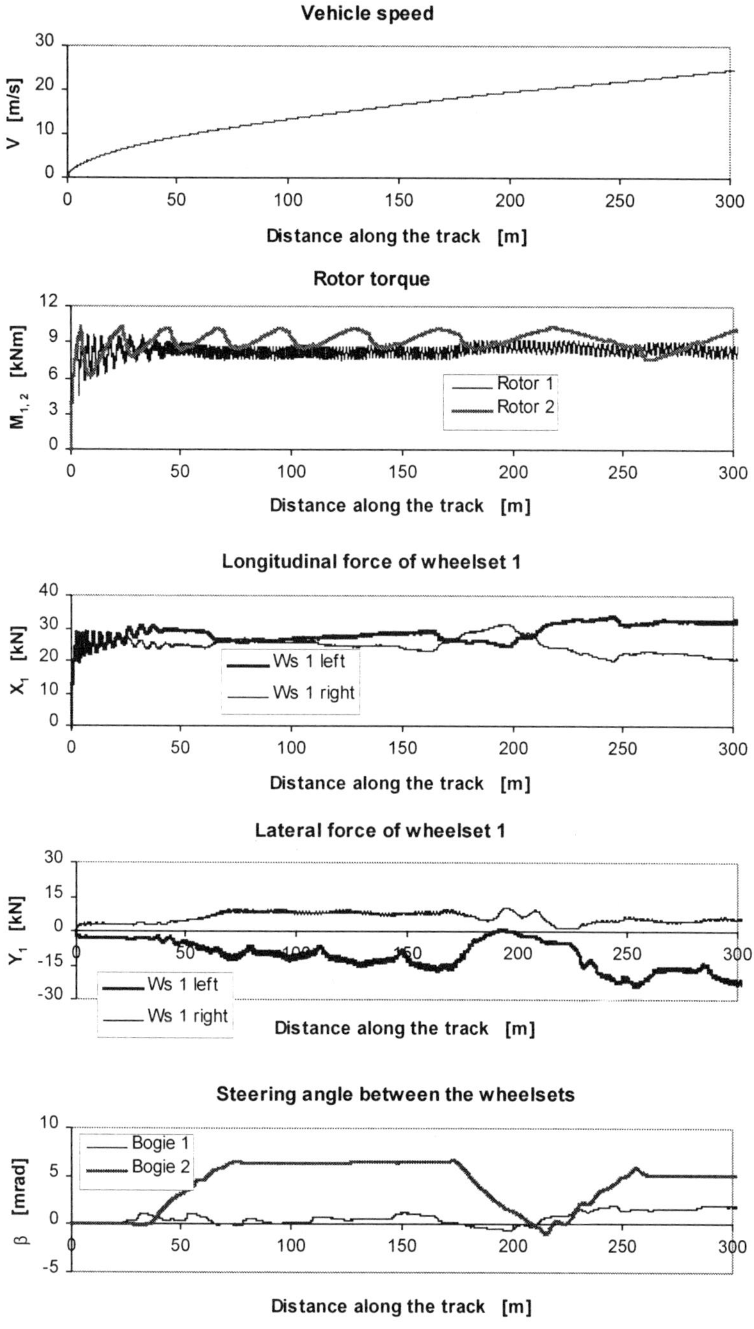

Fig. 11. Starting and acceleration of a locomotive under difficult adhesion conditions on curved track resultant from ADAMS/Rail and MATLAB co-simulation.

5. CONCLUSION

To investigate a mechatronic system of railway traction vehicles, e. g. influence of traction control and vehicle dynamics on the forces between wheel and rail, an adaptation of wheel-rail forces model is necessary. A fast method for the computation of wheel-rail forces developed by the author was extended to simulate various wheel-rail adhesion conditions. The proposed method is suitable for the investigations of problems regarding the traction control and axle drive dynamics and their interaction with vehicle dynamic behaviour. The results calculated using the proposed method show good agreement with measurements and confirm this as a possible way for the simulation of complex mechatronic systems of railway vehicles.

REFERENCES

1. Engel, B., Beck, H. P., Alders, J.: Verschleißreduzierende Radschlupfregelung mit hoher Kraftschlussausnutzung. Elektrische Bahnen, Vol. 96 (1998), No. 6, pp.201-209
2. Kalker, J. J.: On the rolling contact of two elastic bodies in the presence of dry friction. Thesis, Delft 1967
3. Kalker, J. J.: A fast algorithm for the simplified theory of rolling contact. Vehicle System Dynamics 11(1982), pp. 1-13
4. Kalker, J. J.: Über die Mechanik des Kontaktes zwischen Rad und Schiene. ZEV-Glasers Annalen 102 (1978), pp. 214-218
5. Kalker, J. J.: Three-dimensional elastic bodies in rolling contact. Kluwer Academic Publishers, Dordrecht, Netherlands 1990
6. Polach, O.: On problems of rolling contact between wheel and rail. Archives of Transport. Polish Academy of Sciences, Vol. 2 (1990), No. 4, pp. 311-330
7. Polach, O.: Solution of wheel-rail contact forces suitable for calculation of rail vehicle dynamics. Proc. of the 2nd Int. Conference on Railway Bogies, Budapest, Sept. 14-16, 1992, pp. 10-17
8. Polach, O.: A Fast Wheel-Rail Forces Calculation Computer Code. Proc. of the 16th IAVSD Symposium, Pretoria, August 1999, Vehicle System Dynamics Supplement 33 (1999), pp. 728-739
9. Rabinowicz, E.: Friction and wear of materials. John Wiley and Sons, New York, 1965
10. Schreiber, R., Kögel, R.: Identifikationsmethode zur Bestimmung der Adhäsion zwischen Rad und Schiene. ZEV+DET Glasers Annalen, Vol. 120 (1996), No. 2, pp. 48-54
11. Shen, Z. Y., Hedrick, J. K. and Elkins, J. A.: A comparison of alternative creep force models for rail vehicle dynamic analysis. Proc. of the 8th IAVSD-Symposium, Cambridge, MA, August 15-19, 1983, pp. 591-605
12. Vermeulen, P. J., Johnson, K. L.: Contact of non spherical bodies transmitting tangential forces. Journal of Applied Mechanics, Vol. 31 (1964), pp. 338-340
13. Müller, R.: Veränderungen von Radlaufflächen im Betriebseinsatz und deren Auswirkungen auf das Fahrzeugverhalten (Teil 1). ZEV+DET Glasers Annalen, Vol. 122 (1998), No. 11, pp. 675-688

Vehicle System Dynamics Supplement 35 (2001), pp. 23-40

Active Controlled Rail Vehicles for Improved Curving Performance and Response to Track Irregularity

By Yoshihiro SUDA*, Takefumi MIYAMOTO* and Norihiko KATOH*

SUMMARY

To improve the curving performance of railway vehicles is very important for urban transit because tight curve radii are unavoidable in densely populated areas. Poor steering ability on tight curves cause severe wear of wheel flange and railhead, and corrugation on the inner rail. The authors proposed active control to improve the steering ability and response to track irregularities. Theoretical analysis was performed to determine control strategies. A full vehicle model was used for calculation and a scale model with a special single-axle bogie was for experiments on running test bench in this study. The test results show that the proposed control method has suitable performance for curving and in track irregularity response.

1. INTRODUCTION

Rail vehicles normally have self-steering ability on curved tracks. This function is gained from the wheel conicity and rigid connection of two wheels by an axle. This mechanism is very attractive but has the following two problems. One is hunting which is a self-excited vibration of the wheelset under the high-speed condition. The compatibility between stability and curving performance is the main subject to be solved for bogie design. The other is the limitation of the curving ability by the self-steering mechanism. It is not effective under a very tight

* Institute of Industrial Science, University of Tokyo
4-6-1, Komaba, Meguro-ku, Tokyo 153-8505 Japan
Tel: +81-3-5452-6193
Fax: +81-3-5452-6194
E-mail: suda@iis.u-tokyo.ac.jp

curve radius condition with flange contact.

To solve the first problem, and get the compatibility between high-speed stability and curving ability, there are a number of studies and developments promoted in the past. One of the previous studies by the authors is the development of self-steering truck with asymmetric suspension fore-and-aft [1, 2]. For two wheelset bogies, soft stiffness was proposed for the leading axle and hard stiffness for the trailing axle. This self-steering truck has theoretically perfect steering ability if the stiffness is appropriately set. The proposed self-steering truck was adopted for the express train "Shinano" for Central Japan Railway [3, 4]. To achieve soft and hard stiffness for the truck, a pneumatic switching device was used. The control device has already been used for this truck. The measured data shows that the lateral force and wear of wheel and rail have dramatically been reduced.

However, more improvement of the curving performance is necessary for tight curve sections. To improve the curving performance of railway vehicles is very important for urban transit because tight curve radii are unavoidable in densely populated areas. Poor steering ability on tight curves cause severe wear in wheel flange and railhead, and corrugation on the inner rail. Moreover, poor curving ability has the possibility to cause derailment.

Another objective to improve the ability of truck is to obtain better travelling performance in track irregularity sections, i.e., to improve car response for track irregularities. The track and rail maintenance to keep high quality condition needs enormous amount of costs. To achieve the compatibility between these requirements, an option is to use active controlled rail vehicles. Recently there have been several studies of active control for rail vehicles [5-8]. Universal solutions have not been obtained so far, however, for the best truck type or strategy. Moreover, most of the previous studies were made from the theoretical viewpoint. Few studies have implemented experiments or tests with real vehicles.

This paper proposes an active control to improve the steering ability and response to track irregularities. Theoretical analysis was made to determine the truck type for active control and control strategies. A full vehicle model was used for calculation and a scale model with a special single-axle bogie was for experiments on running test bench in this study to verify the proposed concept.

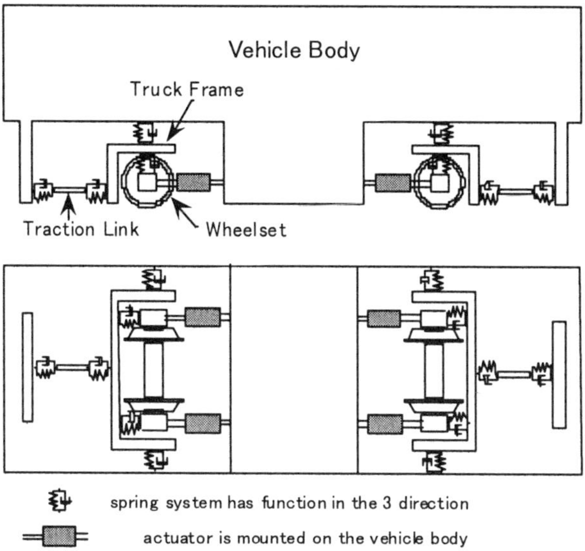

Fig.1 Vehicle model with actuators for yaw motion control

1. CONCEPT OF ACTIVE CONTROL AND VEHICLE MODEL

Two-axle bogies are used for normal commuter rail vehicles. In contrast, the bogie for the proposed active controlled rail vehicles is assumed to be a single-axle bogie. A single axle bogie has several merits to make vehicle lightweight, and wide, for example. Moreover, it is easy to equip actuators in the bogie because of simple truck structure and possible to reduce the number of actuators. The concept of the proposed active controlled vehicle is shown in Fig. 1.

There are some possibilities for the locations of actuators and points to apply the driving force. In this study, we propose that the actuator be installed between the car body and axle box to control the yaw motion of the wheelset. This type of control method is the most effective way, when compared with the cases of actuators installed between car body and truck frame, truck flame and wheelset, and car body and wheelset.

To detect vehicle conditions, the following two methods are assumed. One is the real time measurement of the lateral and yaw displacements between wheel and rail, and the vehicle speed by sensors. For the other method we assume that track information such as curve radius can be known in advance, which can be checked by signals or GPS. For tilting control of the tilting trains in Japan track information known in advance is already used [4].

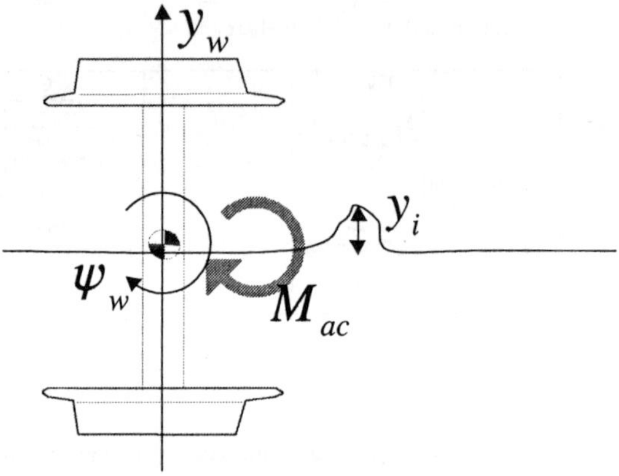

Fig. 2 Control concept for track irregulaliti

2. IMPROVEMENT OF THE RESPONSE TO TRACK IRREGULARITIES

3.1 Control strategy

This section describes the improvement of the ability for response to track irregularities by active control with an actuator between leading wheelset and car body. The basic concept is to apply a yaw moment by sensing the track lateral irregularity as shown in Fig. 2. The equations of motion of a single wheelset for lateral and yawing motions are expressed by Equations (1), where m_w: mass of wheelset, i_w: yaw radius of moment of inertia, k_{wy}: lateral primary stiffness, k_{wx}: longitudinal primary stiffness, V: vehicle speed, b_x: half spacing of primary stiffness, b: half of gauge, r_0: wheel radius, M_{ac}: actuator moment, y_w: wheel lateral displacement, y_i: track irregularity, Ψ_w: yaw displacement of wheelset, and

κ_{11}, κ_{22}:creep coefficient of longitudinal and lateral.

$$m_w \ddot{y}_w + k_{wy} y_w = -2\kappa_{22}\left(\frac{\dot{y}_w}{V} - \psi_w\right)$$

$$m_w i_w^2 \psi_w + k_{wx} b_x^2 \psi_w = -2b\kappa_{11}\left\{\frac{\gamma}{r_0}(y_w - y_i) + \frac{b}{V}\psi_w\right\} + M_{ac} \qquad (1)$$

If we add the actuator moment expressed by Equation (2), the effect of the track irregularity y_i is cancelled. The value of the track irregularity can be estimated form the measurement of the lateral distance between wheel and rail.

$$M_{ac} = \frac{2b\kappa_{11}\gamma}{r_0} \cdot (y_w - y_i) \qquad (2)$$

2.2 Simulation Results

In order to establish a control strategy, computer simulations were made by using multi-body software A'GEM [9] which was originally developed by Queen's University, Kingston, Ontario, Canada. We modified the software in order to treat track irregularities and active controlled force. Figure 3 shows the input track irregularity. The wavelength is 15m and the wave height is 5mm. The vehicle speed is set at 54 km/h.

Figure 4 and 5 show the results of computer simulation, or the attack angle and actuator force. When the results is not controlled, the amplitude of the attack angle is large and its oscillation continues for a relatively long distance. When the active control is in operation, on the other hand, the fluctuation caused by track irregularities can be reduced.

In general, single-axle bogies are sensitive for track irregularities. The use of active control is especially effective in this case.

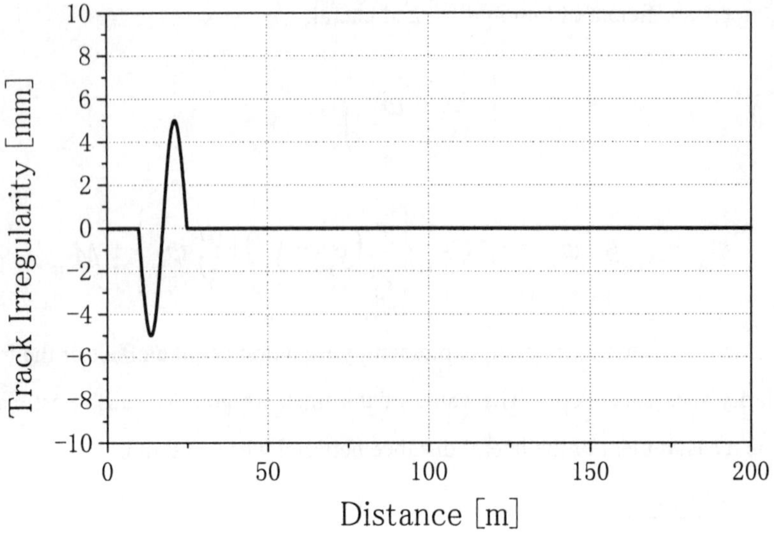

Fig. 3 Input track irregularity for computer simulation.

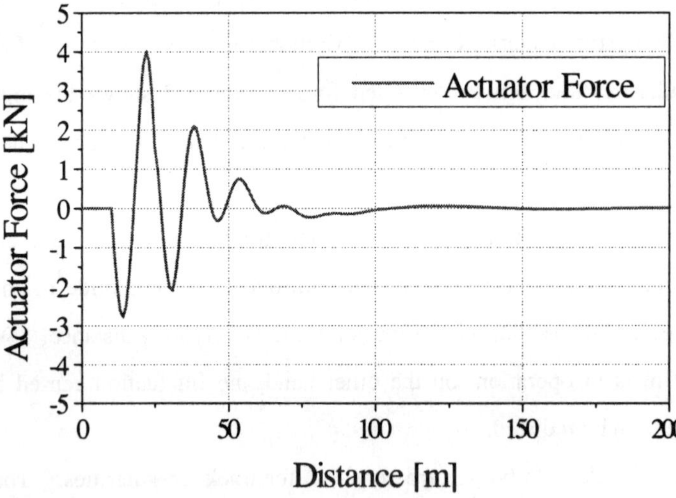

Fig. 4 Actuator force

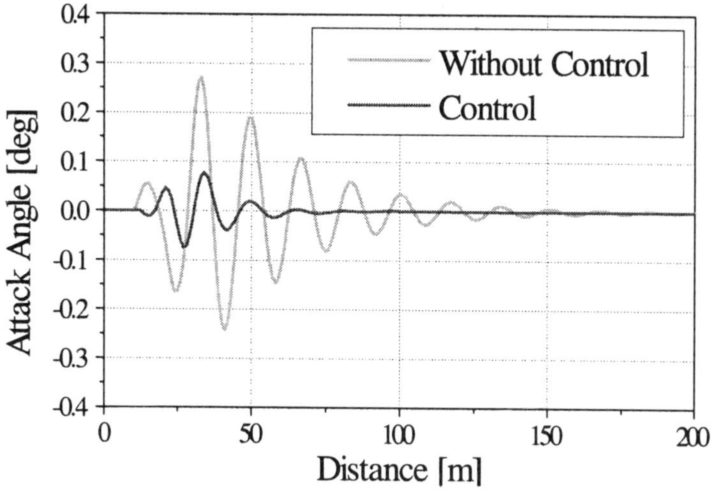

Fig. 5 Simulation results of attack angle in track irregulality section.

3. CURVING PERFORMANCE

4.1 Theoretical Analysis of Curving Performance

To design a control strategy for improvement of the curving performance, we used a simple model of railway vehicle. Because rail cars normally have multiple degrees of freedom and are difficult to treat by a symbolical approach in theoretical studies. Therefore, we adopted an equivalent stiffness model to use shear stiffness and bending stiffness for a two-axle bogie. When only the relative motion of each wheelset is considered, the truck characteristics are expressed by only two equivalent stiffness parameters [10].

A full car model has the primary stiffness and secondary suspension. Under the assumption of steady state condition, it is possible to ignore the mass of bogie frame. Then, using the concept of equivalent stiffness and combined stiffness of primary and secondary suspension, we can obtain a very simple car model for a single-axle bogie rail car as shown in Fig. 6.

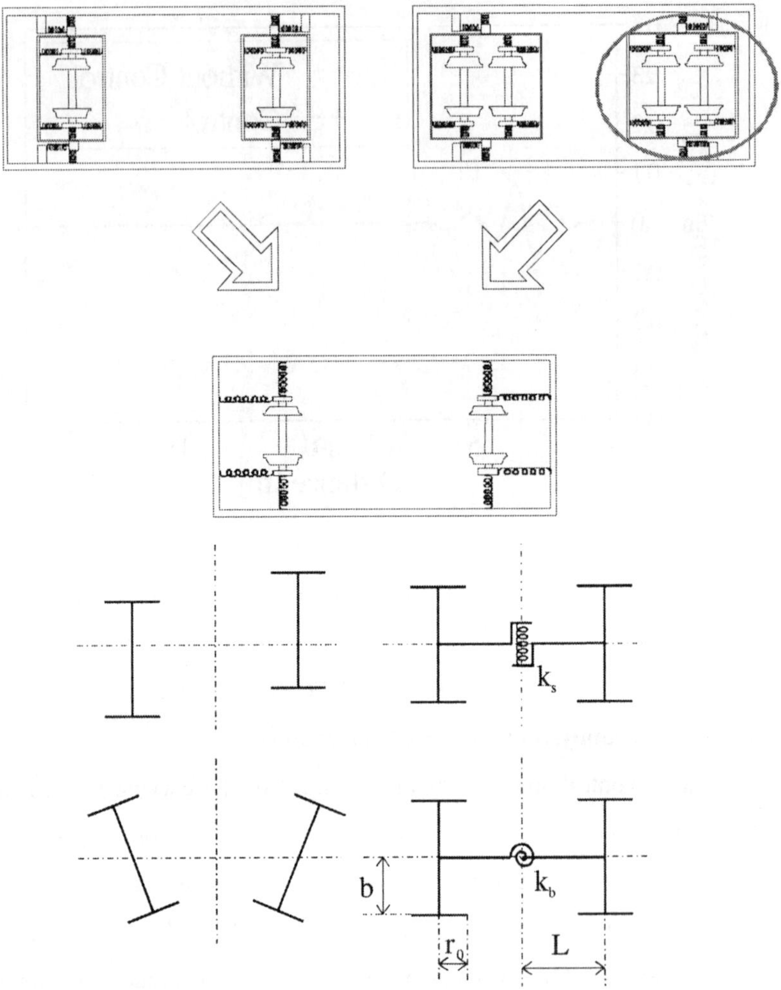

Fig. 6 Simple vehicle model for theoretical study using equivalent stiffness.

4.2 Control Strategy

By the use of the simple equivalent stiffness model, the equations of motion of a car in steady state curving is expressed by Equations (3) [1], where ***M***: inertia matrix, ***B***: damping matrix by creep force, ***C***: damping matrix by dampers, ***E***: creep force matrix, ***K***: stiffness matrix, ***y***: displacement vector, *a*: half of wheel base, and *R*: curve radius. In this Equation, the actuator force M_{ac} is added as a yaw moment of the leading axle. For the purpose simplicity in the analysis of performance, it is enough to control only the leading axle but not the trailing axle.

$$\mathbf{M}\ddot{\mathbf{y}} + \left(\mathbf{C} + \frac{\mathbf{B}}{V}\right)\dot{\mathbf{y}} + (\mathbf{E} + \mathbf{K})\mathbf{y} = \mathbf{K}\mathbf{y}_0 + \mathbf{F}_{\mathbf{act}}$$
$$\mathbf{F}_{\mathbf{act}} = (\mathbf{E} + \mathbf{K})\mathbf{y} - \mathbf{K}\mathbf{y}_0$$
$$\mathbf{F}_{\mathbf{act}} = [0 \quad M_{ac} \quad 0 \quad 0]$$
$$\mathbf{y}_0 = \left[0 \quad -\frac{a}{R} \quad 0 \quad \frac{a}{R}\right]^t \tag{3}$$

Under the assumption of the steady state, the required actuator force can be solved as expressed by Equations (4) and (5). In this study, two separate conditions are considered. One is the off-flange condition, and the other is the on-flange condition. In a moderate curve section, it is possible to negotiate the curve without flange contact if the tread profile is well designed. In this case, the car must be controlled to achieve the zero attack angle condition, or the perfect steering condition. In the ideal condition, the zero attack angle means the zero lateral force acting between rail and wheel.

$$M_{ac} = \frac{2Lk_b}{R} \tag{4}$$

In this Equation, k_b is the equivalent bending stiffness of the single-axle bogie, which takes into consideration both the primary and secondary suspensions.

On a very tight curve section, however, flange contact usually occurs. In this study, the improvement of the curving performance for the tight curve radius condition is one of the major objectives. To avoid flange contact in active control, it is possible in theory to apply the control moment as shown in Equation (5).

$$M_{ac} = \frac{4(-y_f + y_0)bR\gamma\kappa_{11}(k_b k_s r_0 + 4b\gamma\kappa_{11}\kappa_{22})}{Rr_0\{k_b k_s r_0 + 2(ak_s r_0 4b\gamma\kappa_{11})\kappa_{22}\}} + \frac{4ak_b r_0(ak_s r_0 2b\gamma\kappa_{11})\kappa_{22}}{Rr_0\{k_b k_s r_0 + 2(ak_s r_0 4b\gamma\kappa_{11})\kappa_{22}\}}a \tag{5}$$

The control force from the theoretical analysis mentioned above is calculated as shown in Fig. 7. On a moderate curve track, the actuator force is not so large. As a specification in the case considered in this study, which is a commuter train and a curve radius of less than 300m, usual passive trains must have flange contact. To avoid flange contact, a moderate large actuator force is required. This means that an actuator capacity of 30kN force is suitable to negotiate curves without additional lateral forces even for a tight curve radius of 150m.

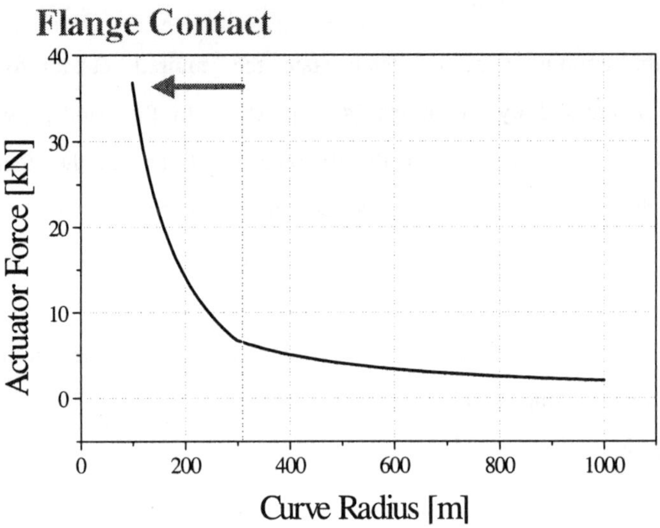

Fig. 7 Actuator force for active steering

4.3 Computer Simulations

Numerical simulations were made by using A'GEM based on the non-linear characteristics of creep forces between wheel and rail. Figure 8 shows the simulation condition. A 200 m curve radius is considered in this calculation. The vehicle speed is 42 km/h and the cant value is set to balance with the centrifugal force.

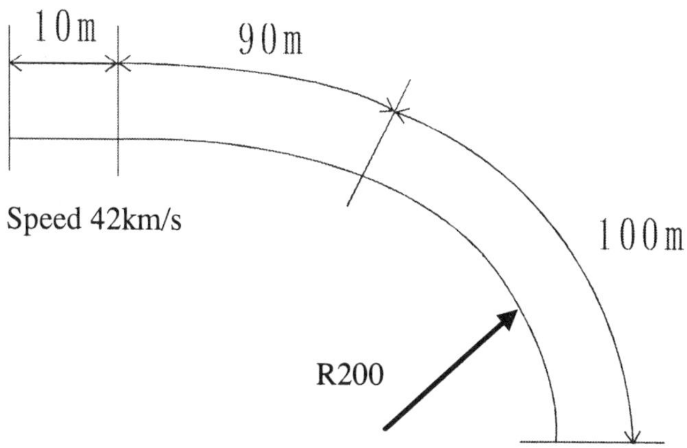

Fig. 8 Track condition of curving simulation.

The results of the computer simulation are shown in Fig. 9-11. The lateral force acting on the leading outer wheel and trailing outer wheel are shown in Figs. 9 and 10, respectively. The dotted line shows the case where the wheelsets is not controlled. In a transition curve section, the lateral force of the leading axle suddenly becomes large when the wheelsets is not controlled. The trailing axle does not show such a large lateral force. The two solid lines show the case of active control. The difference between two cases is in the actuator force. A value of 13kN is gained from theoretical calculation.

In both cases, the lateral force of the leading axle in tight curve sections can dramatically be reduced. In this calculation, only the leading wheelset is controlled. The result shows that sufficient performance is obtained by the active control of only the leading wheelset.

Figure 11 shows the simulation result of attack angle of the leading wheelset. When the active control is not applied, a large attack angle can be seen. By the active control, however, to the zero attack angle was successfully realized.

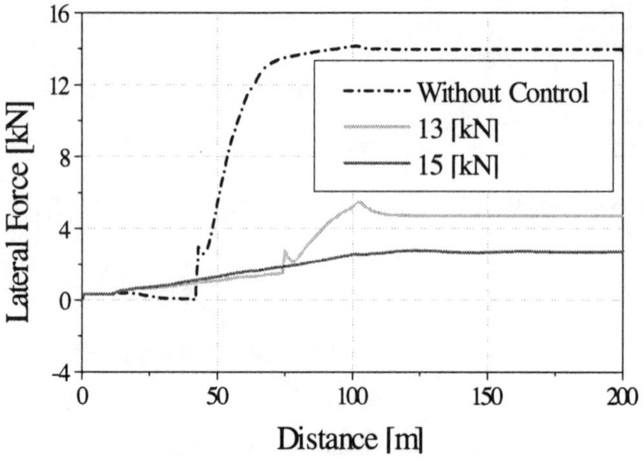

Fig. 9 Lateral force of leading outer wheel

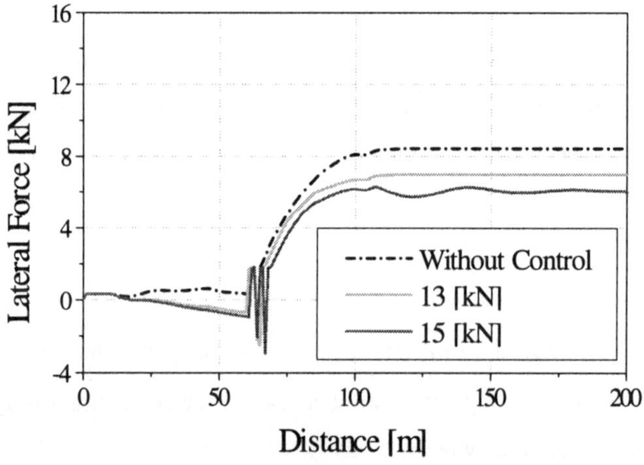

Fig. 10 Lateral force of trailing outer wheel

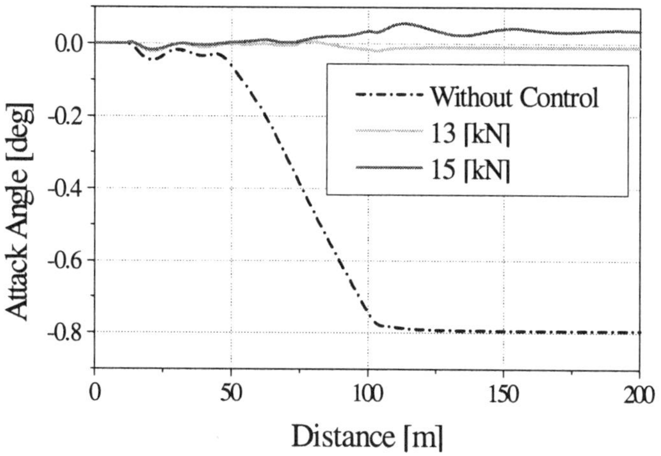

Fig. 11 Attack angle of leading axle

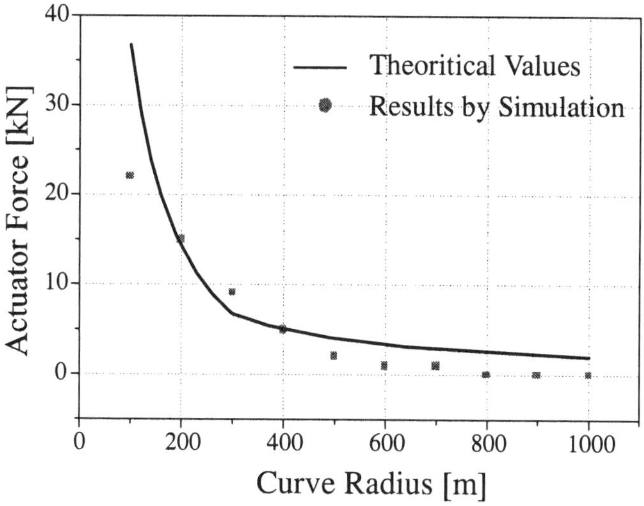

Fig. 12 Actuator force required for the leading axle for steering on curve

An active force of 13kN is gained from the theoretical analysis. In this condition, lateral force can not be canceled completely. If we wish to get a minimum lateral force solution by active control, we must apply an actuator force of 15kN, which is a little larger than theoretical value. In this case, the attack angle shows an over-steering condition.

Figure 12 compares the theoretical value and simulation results for the required actuator force. The solid line is calculated from theoretical analysis (Equation (5)). In the Figure, experimental results are also plotted. A good agreement can be seen. From these results, it is possible to estimate the required actuator force by using simple linear calculations.

Fig. 13 Photo of the model vehicle

4. EXPERIMENTS BY A SCALE MODEL

Experimental investigations were also made in this study. It is important to confirm the reality for the new concepts of the active control by experiments. A special experimental setup with a scale model of 1/10 was designed and used for this study [11]. Figure 13 shows a photo of the model vehicle and a mostly straight wave track section. Two wheelsets and three types of track irregularities were considered to examine the influence of the wavelength of wheelset hunting on the controlled motion. The vehicle model was driven by a special drive unit. A 4m track was available for the measurements. To measure the lateral distance between wheel and rail, two laser sensors were used as shown in Fig. 14.

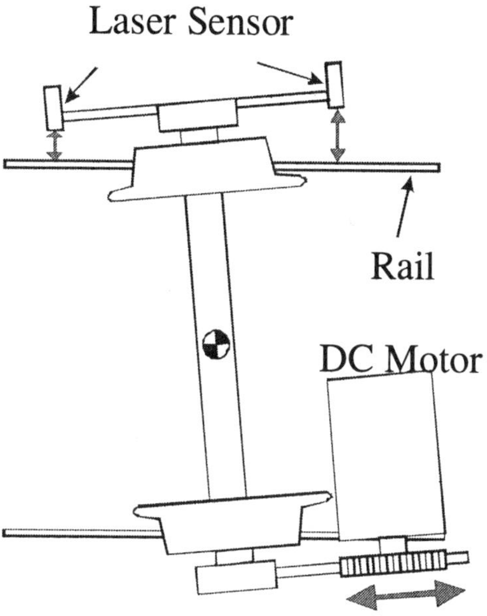

Fig. 14 Control and sensing methods

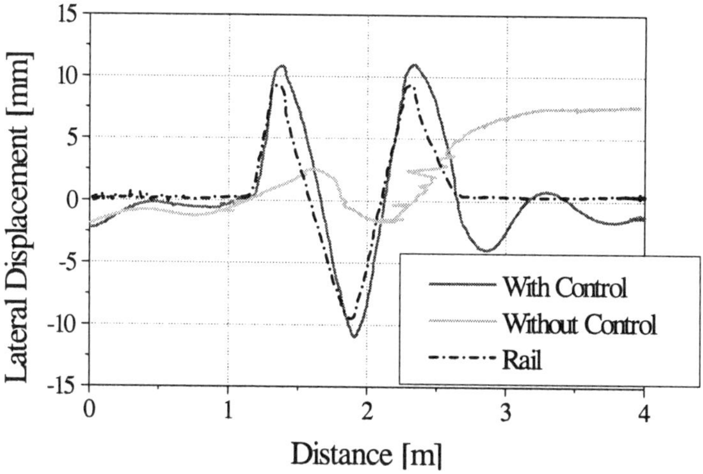

Fig. 15 Experimental results of lateral displacement of wheelset.

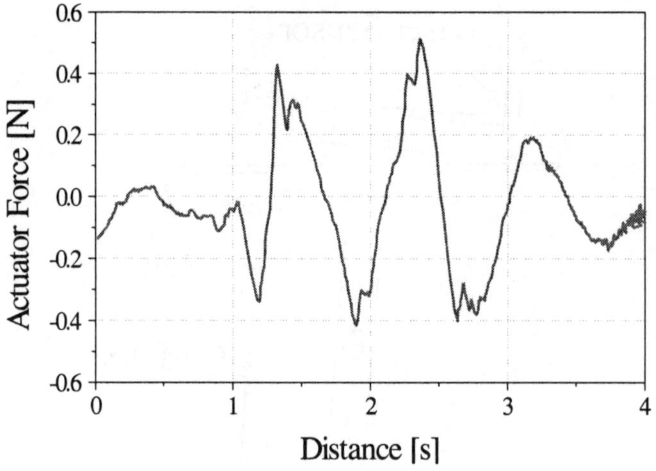

Fig. 16 Experimental results of actuator force

The electric motor (DC motor) used for the actuator can give the yaw moment with feedback control by using the information on the lateral displacement of the wheel. In this study, curving experiments were made by using a track wave shape section.

The experimental results of lateral wheel displacement and actuator force are shown in Figs. 15 and 16. The vehicle speed was 1 m/s. The actuator force was calculated from the information collected, by sensors. The objective of control in this experiment was to track the wheel along the rail shape shown by the dotted line. When active control not applied, tracing the wheelset can not match the rail shape. However, with active yaw control, tracking is achieved by the active control.

5. CONCLUSIONS

In this paper, an active controlled rail vehicle with single-axle bogies was examined to improve the curving performance. The improvement of the response to track irregularities was also considered. The control strategy was examined by theoretical analysis with a simple equivalent stiffness model. The following results were obtained:

- It is effective to apply feedback control added with yaw moment by sensing the lateral displacement between wheel and rail to the leading wheelset only.
- It is proved that the proposed active yaw control of wheelset has good performance. This was done by computer simulation of non-liner characteristics with a full car model and experiments with a scale model.

Acknowledgements

The authors appreciate the Program for Promoting Fundamental Transport Research for financial support.

References

1) Suda, Y., High Speed Stability and Curving Performance of Longitudinally Unsymmetric Trucks with Semi-active Control, ***Vehicle System Dynamics***, 23 (1994), 29.

2) Y. SUDA, R. J. ANDERSON, Improvement of Dynamic Performance of Trucks with Longitudinally Unsymmentiric Structures by Semi-Active Control for Rail Vehicles, ***JSME International Journal***, Series C, 37 (1994), 3, 542.

3) Hashimoto, K. Yamada, T. Metoki, Y. Suda, Development of Unsymmetric Self-Steering Truck Using Soft-Hard Axle Springs, IMechE Conference Transactions International Conference on Better Journey time-Better Business, STECH 96, Birmingham, UK, 1996.

4) Yoshikawa, K., Faster through the curves on JR-Central, ***Railway Gazette International***, 1997-8, 531.

5) Wickens, A. H., Nagy, Z., Guidance and Stability of Articulated Railway Vheicles with Single Axles and Various Forms of Passive and Active Steering, ***Vehicle System Dynamics Supplement***, 33 (1999), 306.

6) Gretzschel, M., Bose, L., A Mechatronic Approach for Active Influence on Railway Vehicle Running Behaviour, ***Vehicle System Dynamics Supplement***, 33 (1999), 418.

7) Powell, A., Mechatronic Control of an Actively Guided Rail Vehicle, ***Vehicle System Dynamics Supplement***, 33 (1999), 442.

8) Mei, T. X., Goodall, R. M., Wheelset Control Strategies for a Two-Axle Railway Vehicle, ***Vehicle System Dynamics Supplement***, 33 (1999), 653.

9) Anderson, R. J., 1990, A'GEM Rail Vehicle Dynamics Software Package Users Manual.

10) Wickens, A. H., Steering and Dynamic Stability of Railway Vehicles, ***Vehicle System Dynamics***, 5 (1975), 15.

11) Miyamoto, T., Suda, Y., Katoh, N., Komine, H., Experiments on Vehicle Dynamics of Wheel/Rail System with Scale Model, Proceeding of the 8th Transportation and Logistics Conference, JSME, (1999), No.99-59, 219 (in Japanese).

Vehicle System Dynamics Supplement 35 (2001), pp. 41-55

Stabilization Control for the Hunting Motion of a Railway Wheelset

H. YABUNO*, T. OKAMOTO* AND N. AOSHIMA*

ABSTRACT

Rail vehicles experience the problem of hunting above a certain speed. The phenomenon is kind of self-excited oscillation due to the unsymmetrical stiffness matrix in the equation of motion. In the present paper, we propose a stabilization control method for the hunting motion in a wheelset model with two degrees of freedom by reducing the unsymmetrical effect of the stiffness matrix. It is theoretically shown under this control that the critical speed is infinite and no hunting motion occurs at any finite forward speed. Furthermore, an experimental study is conducted using a simple model and the validity of the control method is experimentally discussed.

Key words: bifurcation, circulatory system, critical speed, hunting motion, stabilization control, unsymmetrical stiffness matrix, vehicle dynamics, wheelset.

1 INTRODUCTION

Rail vehicles experience the problem of hunting above a critical forward speed. The phenomenon is induced by a non-conservative contact force (for example, see [1]) acting between wheels and rails, and is a flutter-type self-excited oscillation [2, 3] due to the unsymmetrical stiffness matrix. Recently, nonlinear characteristics have become better understood as a result of advances in the bifurcation theory (for example, see [4]). It was shown by bifurcation analysis that the wheelset undergoes Hopf bifurcation at a critical speed (a comprehensive review is given in [5]). It was also clarified that the hunting motion can occur below the critical speed obtained by the linear theory, depending on the magnitude of disturbance, in the case when the nonlinear characteristics of the bifurcation are subcritical [3, 6]. The subcritical Hopf bifurcation is transformed into a supercritical one by bifurcation control [7] and the stabilization is carried out independent of the magnitude of disturbance for a wheelset model.

*Institute of Engineering Mechanics and Systems, University of Tsukuba, Tsukuba City, 305-8573, Japan.

In the present paper, we propose a stabilization control method for the hunting motion in a simple wheelset model. We focus on the principal cause of the hunting motion, which is asymmetry in the stiffness matrix of the equation of motion. In order to reduce the effect of the unsymmetrical component of the matrix, a lateral force proportional to yawing motion is applied and then it is shown that the critical speed becomes infinite. Consequently, a bifurcation-free dynamics is theoretically realized and the wheelset is stabilized at any forward speed independent of the disturbance. An experimental study is also conducted using a simple model and the validity of the proposed control method is discussed by comparing with theoretical results.

2 ANALYTICAL MODEL AND EQUATION OF MOTION

We consider the behavior of a single suspended railway wheelset at a constant forward speed, v, on a straight track. The mathematical model has two degrees of freedom: lateral motion, y, and yawing motion, ψ, as shown in Fig. 1, where r_0, k_x, γ_e, and d_0 are centered wheel rolling radius, longitudinal stiffness, wheel tread angle, and half of track gauge, respectively.

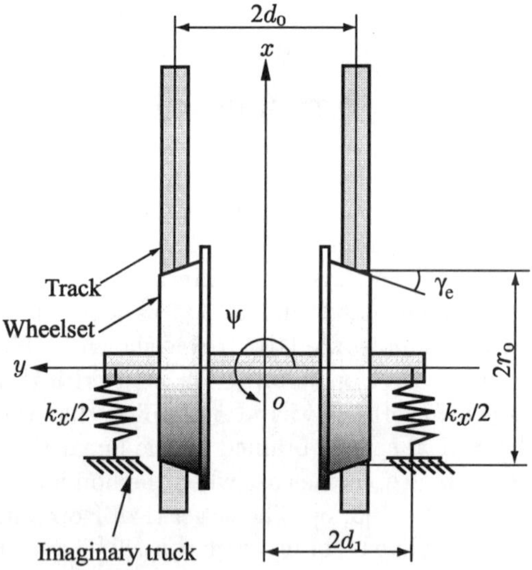

Fig. 1. Wheelset model

Using half of track gauge, d_0, as the representative length and an inverse number of the natural frequency, ω_ψ, as the representative time, the dimensionless equations governing the dimensionless lateral motion, y^*, and yawing motion, ψ, are obtained as follows (see APPENDIX A):

$$\ddot{\boldsymbol{x}} + D\dot{\boldsymbol{x}} + K\boldsymbol{x} = 0 \tag{1}$$

$$\boldsymbol{x} = \begin{bmatrix} y^* \\ \psi \end{bmatrix}, \quad D = \begin{bmatrix} d_{11}/v^* & 0 \\ 0 & d_{22}/v^* \end{bmatrix}, \quad K = \begin{bmatrix} k_{11} & k_{12} \\ k_{21} & k_{22} \end{bmatrix}.$$

The elements of the above matrices are expressed as follows:

$$\begin{aligned} d_{11} &= 2\kappa_y/(md_0\omega_\psi^2), d_{22} = 2\kappa_x d_0/(I\omega_\psi^2), \\ k_{11} &= (1 - l_0/l)(\omega_y/\omega_\psi)^2, k_{22} = 1, \\ k_{12} &= -2\kappa_y/(md_0\omega_\psi^2), k_{21} = 2d_0^2\kappa_x\gamma_e/(Ir_0\omega_\psi^2). \end{aligned}$$

Here, m, I, l, l_0, and v^* are the mass of the wheelset, the moment of inertia, the length of the spring in the equilibrium state, the natural length of the spring, and the dimensionless forward speed, respectively. The asterisk denotes the dimensionless value. κ_x and κ_y are creep coefficients in the x and y directions, respectively. ω_ψ is the natural frequency of the wheelset when its motion in the y-direction is fixed and ω_y is the natural frequency of the wheelset when its motion in the y-direction is liberated assuming that the wheelset is suspended in the y-direction, i.e., the springs in Fig. 1 are relocated along the y-axis. The parameter values corresponding to the experimental apparatus in this study are as follows: $m = 2.13$ kg, $2d_0 = 0.098$ m, $2d_1 = 0.15$ m, $r_0 = 0.036$ m, $l = 0.056$ m, $l_0 = 0.035$ m, $\gamma_e = 1/40$, $\omega_y = 19.0$ rad/s, $\omega_\psi = 35.0$ rad/s, and $\kappa_x = 180$N, and $\kappa_y = 0.8\kappa_x$ (see APPENDIX B). The dimensionless coefficients of Eq. (1) are as follows: $d_{11} = 2.40, d_{22} = 4.15, k_{11} = 0.0758, k_{12} = -2.40, k_{21} = 0.141, k_{22} = 1.00$.

The damping matrix D in Eq. (1) is symmetric, but the stiffness matrix is not symmetric due to the contact force (the creep force). Because the unsymmetrical matrix can be generally and uniquely represented as the sum of a symmetric matrix and a skewsymmetric matrix, the stiffness matrix, K, can be separated as

$$K = K_s + K_u, \tag{2}$$

$$K_s = \frac{1}{2}(K + K^t), \quad K_u = \frac{1}{2}(K - K^t), \tag{3}$$

where K_u is a skewsymmetric matrix and can induce hunting motion due to flutter-type instability. K_s and K_u are concretely expressed as follows:

$$K_s = \begin{bmatrix} k_{s11} & k_{s12} \\ k_{s12} & k_{s22} \end{bmatrix}, \quad K_u = \begin{bmatrix} 0 & k_{u12} \\ -k_{u12} & 0 \end{bmatrix}, \tag{4}$$

$$k_{s11} = (1 - l_0/l)(\omega_y/\omega_\psi)^2, k_{s22} = 1,$$
$$k_{s12} = d_0^2 \kappa_x \gamma_e/(I r_0 \omega_\psi^2) - \kappa_y/(m d_0 \omega_\psi^2),$$
$$k_{u12} = -d_0^2 \kappa_x \gamma_e/(I r_0 \omega_\psi^2) - \kappa_y/(m d_0 \omega_\psi^2).$$

The dimensionless elements of matrices, D, K_s, and K_u, corresponding to the subsequent experimental apparatus are as follows: $d_{11} = 2.40, d_{22} = 4.15, k_{s11} = 0.0758, k_{s12} = -1.13, k_{s22} = 1.00$, and $k_{u12} = -1.27$.

3 STABILITY OF THE WHEELSET

The characteristic equation of Eq. (1) is expressed as

$$\lambda^4 + \frac{d_{22}+d_{11}}{v^*}\lambda^3 + \left(k_{s22} + k_{s11} + \frac{d_{11}d_{22}}{v^{*2}}\right)\lambda^2 + \frac{d_{11}k_{s22}+d_{22}k_{s11}}{v^*}\lambda + k_{s11}k_{s22} + k_{u12}^2 - k_{s12}^2 = 0. \quad (5)$$

The characteristic roots of Eq. (5) are varied along the arrows in Fig. 2, with increasing forward speed, v^*. The speed at which the root loci intersect the imaginary axis corresponds to the critical speed for hunting motion, v_c^*, which is a finite number; above this speed, the hunting motion occurs. At the critical speed, the characteristic roots are a pair of conjugate purely imaginary values and a pair of complex conjugates with a negative real part. Therefore, the wheelset is destabilized through Hopf bifurcation [4] at the critical speed. The objective of this research is to propose a stabilization control method for the hunting motion, in which bifurcation-free dynamics is realized.

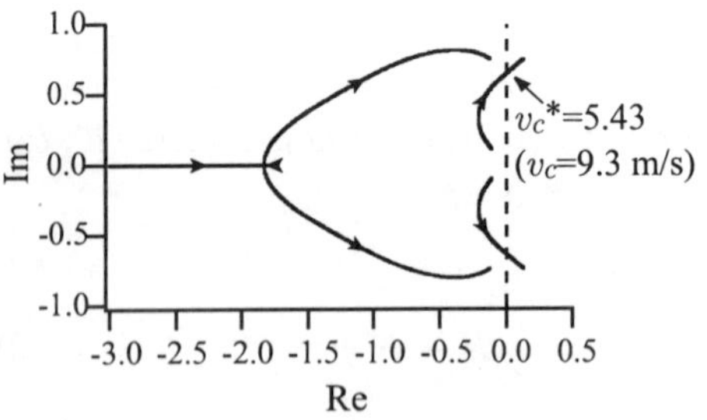

Fig. 2. Root loci (without control)

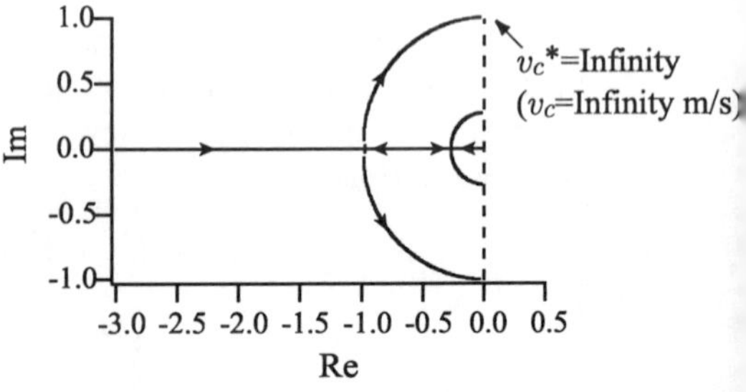

Fig. 3. Root loci (with control)

4 STABILIZATION OF HUNTING MOTION BY MAKING STIFFNESS MATRIX SYMMETRICAL

Hunting motion is produced by the existence of the unsymmetrical component of the stiffness matrix. Consequently, in order to reduce the unsymmetrical effect, we apply dimensionless control force, f^*, in the lateral direction, which is proportional to yawing motion,

$$f^* = -h^*\psi. \tag{6}$$

Then, the skewsymmetric matrix, K_u, is changed to K'_u as follows:

$$K'_u = \begin{bmatrix} 0 & k_{u12} + h^* \\ -k_{u12} & 0 \end{bmatrix}. \tag{7}$$

When we set the feedback gain as $h^* = -2k_{u12}$, the matrix, K'_u and the stiffness matrix, $K = K_s + K'_u$, become completely symmetric. Then, the root loci are drawn, as shown in Fig. 3. These loci qualitatively differ from those of the case without control (Fig. 2); the root loci exist only on the left-half plane at any finite forward speed and all four eigenvalues are purely imaginary numbers at infinite speed. Therefore, no hunting motion occurs at any finite speed. However, in the practical usage of this control method, we cannot set exactly the feedback gain as $h^* = -2k_{u12}$. For this reason, let us examine the relationship between the feedback gain and the critical speed. Under feedback control, the characteristic equation, Eq. (5), is modified as

$$\lambda^4 + \frac{d_{11} + d_{22}}{v^*}\lambda^3 + \left(k_{s11} + k_{s22} + \frac{d_{11}d_{22}}{v^{*2}}\right)\lambda^2 + \frac{d_{11}k_{s22} + d_{22}k_{s11}}{v^*}\lambda + k_{s11}k_{s22} - k_{s12}^2 + k_{u12}^2 + h^*(k_{u12} - k_{s12}) = 0. \tag{8}$$

Furthermore, we obtain the variation of the critical speed with the magnitude of the feedback gain, h^*, as shown in Fig. 4. With increasing h^*, the critical speed increases, and is infinite in a range of the gain (Too large feedback makes the system unstable and, above $h^* > 2.94$, the critical speed is 0, i.e., hunting motion occurs at any low speed).

Next, we examine this range using the characteristic equation. At an infinite speed, $v^* \to \infty$, the characteristic roots satisfy the following equation:

$$\lambda^4 + (k_{s11} + k_{s22})\lambda^2 + k_{s11}k_{s22} - k_{s12}^2 + k_{u12}^2 + h^*(k_{u12} - k_{s12}) = 0. \tag{9}$$

When all roots of the above equation are purely imaginary, i.e., on the imaginary axis in the s plane, the characteristic roots exist in the left-half s plane at any finite speed. Then, the critical speed is infinite and no hunting motion occurs at any finite speed. It can be found from Eq. (9) that all roots are purely

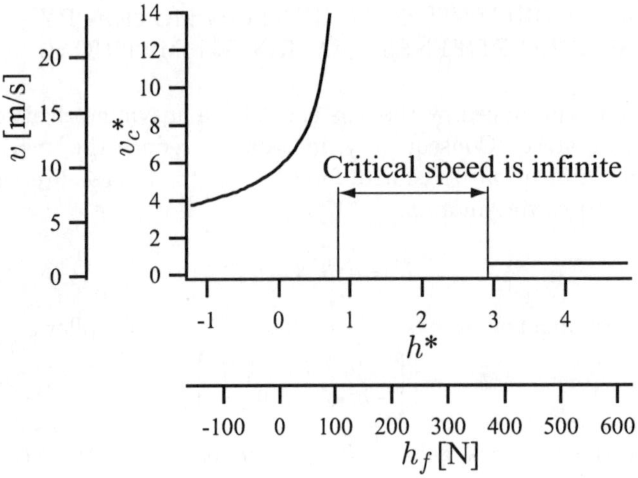

Fig. 4. Relationship between feedback gain and critical speed (theory)

imaginary at infinite speed (the critical speed is infinite) in the following range of h^*:

$$\begin{aligned} &\{k_{s12}^2 - k_{u12}^2 + (k_{s11} - k_{s22})^2/4\}/(k_{u12} - k_{s12}) \\ &< h^* < \\ &(-k_{s12}^2 - k_{u12}^2 - k_{s11}k_{s22})/(k_{u12} - k_{s12}), \end{aligned} \tag{10}$$

This range corresponds to the parameters of the subsequent experimental apparatus is $0.889 < h^* < 2.94$ (the range in the dimensional form, $h = d_0\omega_\psi^2 h^*$, is $53.1\ \mathrm{m/s^2} < h < 176\ \mathrm{m/s^2}$ and the range of $h_f = mh$ is $113\ \mathrm{N} < h_f < 375\ \mathrm{N}$). Then the root loci qualitatively resemble those of Fig. 3 and the critical speed for hunting motion becomes infinite, even if the stiffness matrix cannot be changed into a completely symmetrical one.

Meanwhile, it is also possible to make the critical speed higher than that of the case without control if the control force proportional to the lateral velocity is applied in the lateral direction (the feedback gain is g^*). However, in this control method, the critical speed cannot become infinite under any feedback gain as shown in Fig. 5, because the root loci are kept qualitatively the same as that of the case without control (Fig. 2).

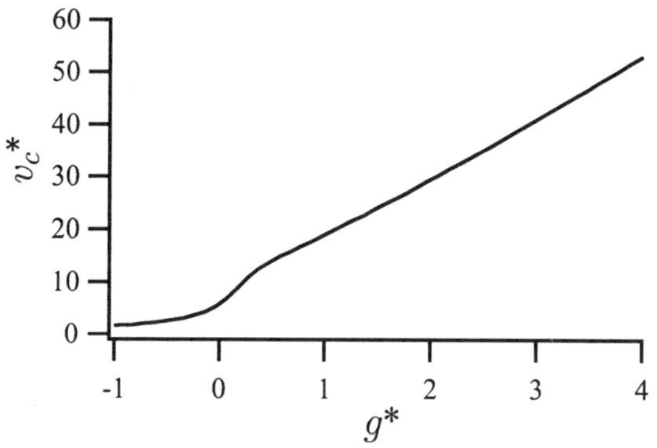

Fig. 5. Relationship between feedback gain and critical speed in the case when the lateral velocity is fed back (theory)

5 EXPERIMENT

In this section, we experimentally investigate the effect of the proposed stabilization control for hunting motion. Photographs of the experimental apparatus are shown in Figs. 6(a) and (b). The wheelset can be freely moved in both y and ψ directions by linear and angular bearing. The radius of the roller rig is 0.15 m. The control force in the y-direction is produced by the linear motor (thrust constant, 2 N/A; maximum thrust, 8 N; frequency response characteristics, 300 Hz) shown in Fig. 6(b). Lateral and yawing motions measured from the displacement sensor and the angular detector are input into a PC. Then, the thrust of the linear motor needed for the control is calculated in real time, according to Eq. (6); the dimensional control force is $f = h_f \psi$. The value is further converted into voltage by considering the gain of the motor driver and the thrust constant number of the motor, and then is output into the motor driver from the PC, where the motor driver can send the current to the motor proportional to the input voltage of the motor driver (refer to Fig. 7). To check the power of the motor needed in the control, the current of the motor can be also monitored in the experiment using a current sensor.

5.1 Hunting motion and its critical speed in the case without control

First, we experimentally determine the critical speed by the following method. We observe the behavior of the wheelset at many forward speeds under the condition where artificial noise does not exist, i.e., when only frustration noise

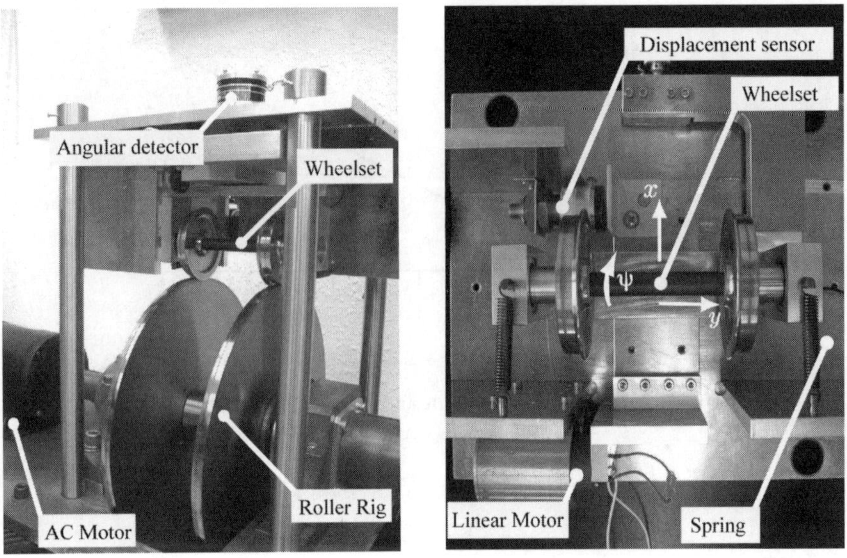

(a) General view (b) Bottom view

Fig. 6. Experimental apparatus

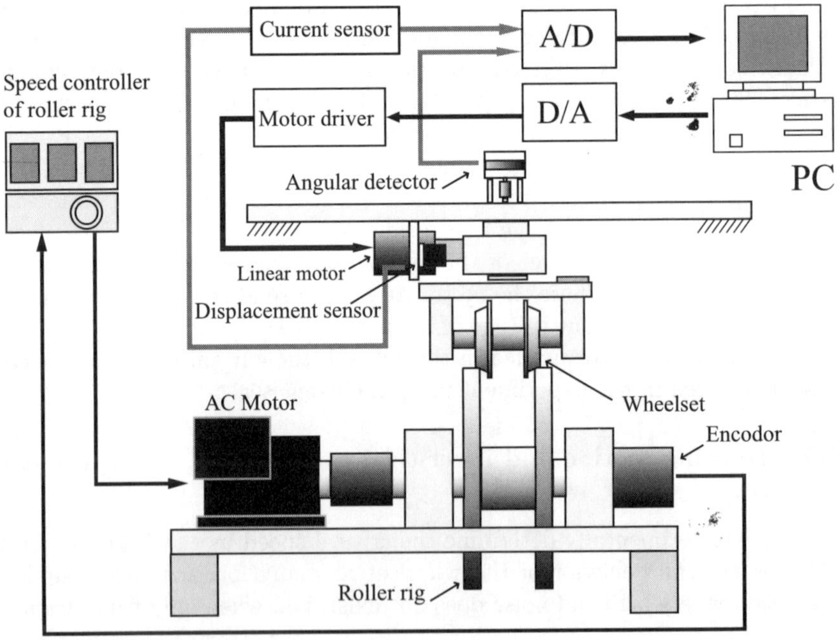

Fig. 7. Control system

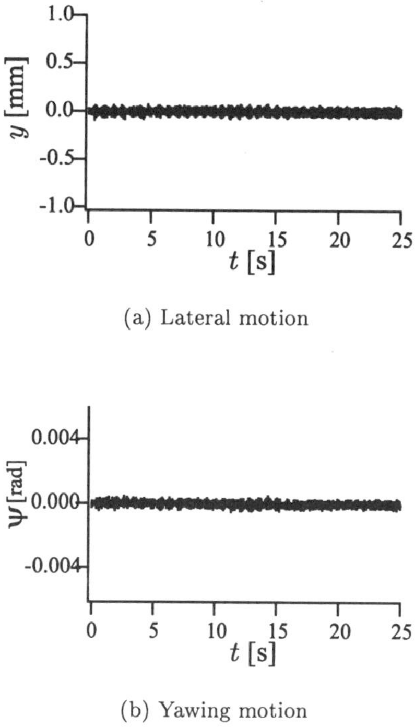

(a) Lateral motion

(b) Yawing motion

Fig. 8. Time history of wheelset motion in case without control (v=9.0 m/s; below the critical speed)

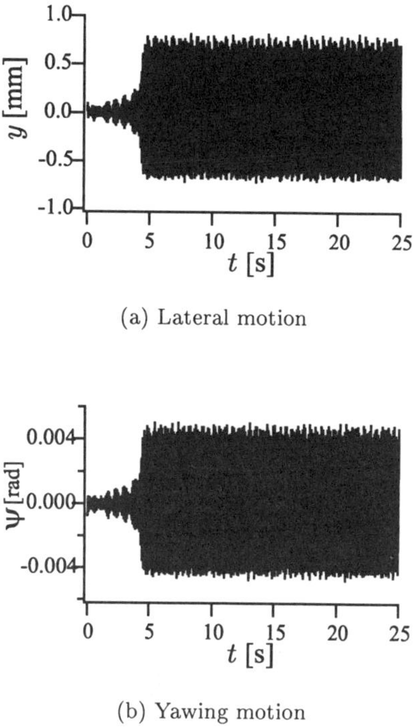

(a) Lateral motion

(b) Yawing motion

Fig. 9. Time history of wheelset motion in case without control (v=10.0 m/s; above the critical speed)

exists. As a result, in the case when the forward speed is less than a certain value, the lateral and yawing oscillations maintain small amplitudes as in Fig. 8, while in the case when the forward speed is more than the value, the lateral and yawing oscillations grow with time, as in Fig. 9, where the growth of the amplitude is limited by flange contact of the wheel to the roller rig. Then, we regard the speed for the boundary of the stability of the motions, as the critical speed, v_c, which is $v_c = 9.3$m/s in the experimental apparatus.

5.2 Stabilization control for hunting motion

In the case when the forward speed is above the critical speed, i.e., $v = 10.0$m/s, we show experimental time histories of lateral and yawing motions and the current of the motor, under the proposed stabilization control ($h = 144$m/s^2,

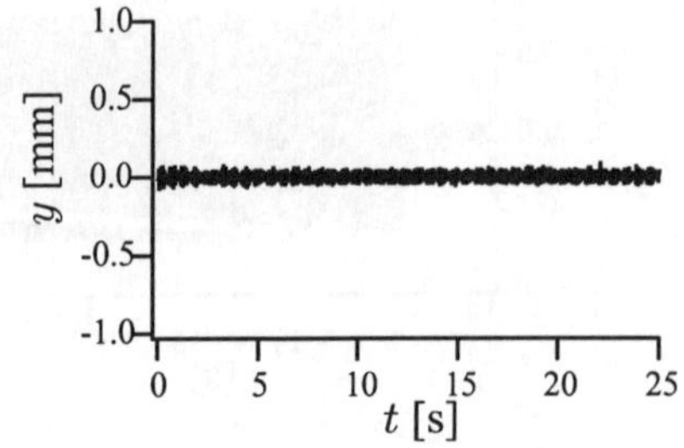

(a) Lateral motion

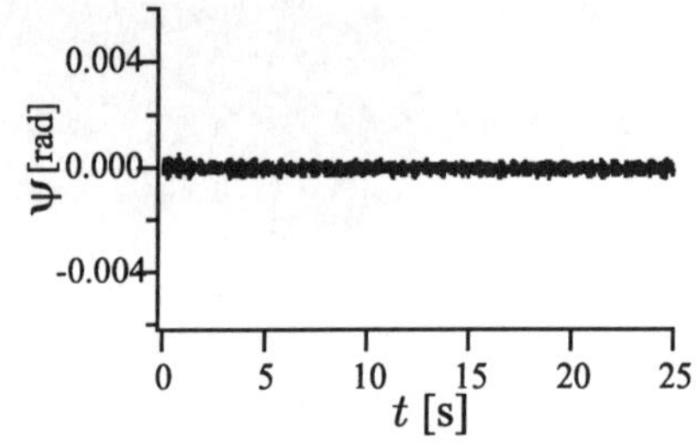

(b) Yawing motion

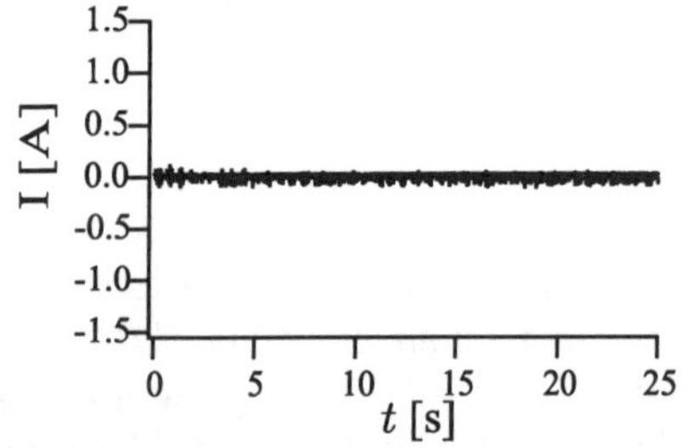

(c) Current of linear motor

Fig. 10. Time history of wheelset motion in case with control (v=10.0 m/s; above the critical speed: without artificial disturbance ($h_f = 307$ N))

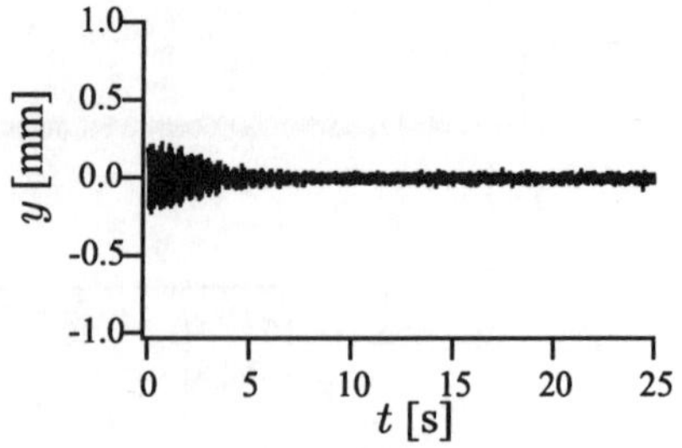

(a) Lateral motion

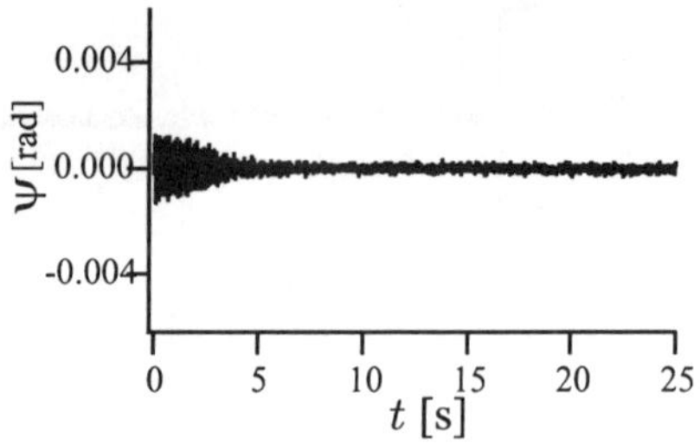

(b) Yawing motion

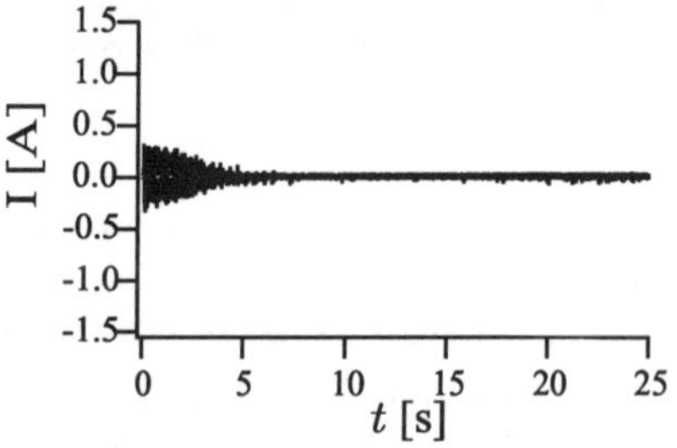

(c) Current of linear motor

Fig. 11. Time history of wheelset motion in case with control (v=10.0 m/s; above the critical speed: with artificial disturbance ($h_f = 307$ N))

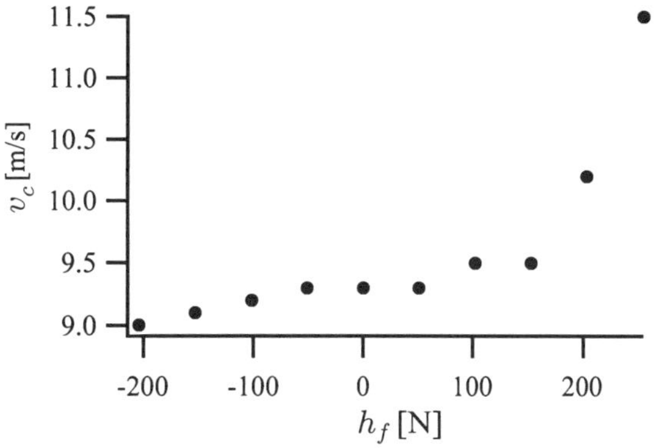

Fig. 12. Relationship between feedback gain and critical speed (experiment)

$h_f = 307$ N) in Fig. 10. By comparing with Fig. 9, stabilization of hunting motion is experimentally confirmed. The small current in the motor is due to the feedback of the small oscillation of ψ caused by the roughness of the roller rig. In other words, little work by the motor is required for stabilization of hunting motion.

Under artificial initial disturbance, the behavior of the wheelset is shown in Fig. 11. It is seen from the time history of the current that only in the transient response, the motor does the work for stabilization of hunting motion. Then, the maximum power produced by the linear motor is 1.75 W, because the resistance of the linear motor is 5.5Ω. In the steady state after the transient state, the current of the motor is very small and the motor does little work. The small current of the motor in the steady state is due to the feedback of the small oscillation of ψ as in the case without initial disturbance.

Finally, we show the experimental relationship between the critical speed and the feedback gain in Fig. 12. This figure is in qualitative agreement with the theoretical one in Fig. 4, but the range of the feedback gain at the infinite critical speed cannot be experimentally detected.

6 CONCLUDING REMARKS

A stabilization control method for hunting motion in a wheelset model with two degrees of freedom is proposed by focusing on the excitation mechanism. The effect of the unsymmetrical stiffness matrix which induces hunting motion is reduced by the lateral feedback control force proportional to the yawing motion, and then it is theoretically shown that the critical speed becomes infinite in a range of feedback gain. By using a small model, it is experimentally confirmed that the critical speed is increased by the proposed control method. It is, however, impossible to detect the experimental range of the feedback gain at infinite critical speed. This seems to be due to the difficulty of modeling of the creep force. To overcome this difficulty, the combination of H_∞ and the proposed method is required. Such a control system will be presented in a future work.

ACKNOWLEDGEMENTS

This work was supported by a Grant-in-Aid for Scientific Research from the Japanese Ministry of Education, Science, Sports and Culture (12650438).

REFERENCES

1. Kalker J. J.: Three-dimensional Elastic Bodies in Rolling Contact. Kluwer, 1990.
2. Wickens, A. H.: The Dynamics Stability of a Simplified Four-wheel Railway Vehicle Having Profiled Wheels. *Int. J. Solids Structures*, 1 (1965), pp. 385-406.
3. Xu, G., Troger, H., and Steindl, A.: Global Analysis of the Loss of Stability of a Special Railway Body. *In W. Schiehlen, (ed), Nonlinear Dynamics in Engineering Systems*, pp. 345–352, Springer-Verlag (Berlin), 1990
4. Guckenheimer, J. and Homes, P.: Nonlinear Oscillations, Dynamical Systems, and Bifurcations of Vector Fields. Springer-Verlag (NewYork), 1983.
5. True, H.: Dynamics of a Rolling Wheelset. *Appl Mech Rev*, 46 (1993), pp. 438-444.
6. Molle, D. and Gash, R.: Nonlinear Bogie Hunting. *Nonlinear Bogie Hunting, Proceedings 7th IAVSD-Symposium (A. H. Wickend ed.) Swets and Zeitlinger B. V., Lisse*, pp 455–467, 1982.
7. Yabuno, H.: Bifurcation Control of Parametrically Excited Duffing System by a Combined Linear-Plus-Nonlinear Feedback Control, *Nonlinear Dynamics*, 12 (1997), pp. 263–274.

7 APPENDIX A

First, we elementally derive creep force on the wheelset. The longitudinal and lateral creepages, v_x and v_y, can be expressed as follows:

$$v_{xl} = \frac{v - d_0 d\psi/dt - r_l\omega}{v} \tag{11}$$

$$v_{xr} = \frac{v + d_0 d\psi/dt - r_r\omega}{v} \tag{12}$$

$$v_{yl} = \frac{-v\psi + dy/dt}{v} \tag{13}$$

$$v_{yr} = \frac{-v\psi + dy/dt}{v}, \tag{14}$$

where v, d_0, and r are forward speed, half of track gauge, and radius of the wheel at the contact point to the tail, respectively. The subscripts, l and r, denote the value with respect to the left wheel and the right one, respectively. By assuming that longitudinal and lateral creep coefficients are constants, κ_x and κ_y, the longitudinal and lateral creep forces are expressed as follows:

$$\begin{aligned} T_{xl} = -\kappa_x v_{xl}, \quad T_{xr} = -\kappa_x v_{xr}, \\ T_{yl} = -\kappa_y v_{yl}, \quad T_{yr} = -\kappa_y v_{yr}. \end{aligned} \tag{15}$$

Consequently, the lateral force, F_{cy}, and the moment with respect to the z axis, M_{cz}, based on creep forces acting on the wheelset are expressed as follows:

$$F_{cy} = T_{yl} + T_{yr} = -2\kappa_y \left(\frac{dy/dt}{v} - \psi \right) \tag{16}$$

$$M_{cz} = d_0(T_{xr} - T_{xl}) = -\frac{2d_0^2\kappa_x}{v}\frac{d\psi}{dt} - \frac{2d_0\kappa_x\gamma_e}{r_0}y, \tag{17}$$

where γ_e and r_0 are the tread angle and the centered wheel rolling radius, respectively.

Considering the restoring force by the longitudinal spring stiffness as shown in Fig. 1, the equations of motion governing the lateral and yawing motions are obtained as follows:

$$m\frac{d^2y}{dt^2} = -\frac{2\kappa_y}{v}\frac{dy}{dt} - k_x\left(1 - \frac{l_0}{l}\right)y + 2\kappa_y\psi \tag{18}$$

$$I\frac{d^2\psi}{dt^2} = -\frac{2d_0^2\kappa_x}{v}\frac{d\psi}{dt} - \frac{2d_0\kappa_x\gamma_e}{r_0}y - k_x d_1^2\psi, \tag{19}$$

where m, I, l, and l_0 are the mass of the wheelset, the moment of inertia, the length of the spring in the equilibrium state, and the actual length of the spring, respectively. Also, the above equations are rewritten as follows:

$$\frac{d^2y}{dt^2} = -\frac{2\kappa_y}{mv}\frac{dy}{dt} - \frac{k_x}{m}\left(1 - \frac{l_0}{l}\right)y + \frac{2\kappa_y}{m}\psi \tag{20}$$

$$\frac{d^2\psi}{dt^2} = -\frac{2d_0^2\kappa_x}{Iv}\frac{d\psi}{dt} - \frac{2d_0\kappa_x\gamma_e}{Ir_0}y - \frac{k_x d_1^2}{I}\psi. \tag{21}$$

Using half of track gauge, d_0, as the representative length and an inverse number of the natural frequency, ω_ψ, as the representative time, the dimensionless equations of the model with respect to dimensionless lateral direction, $y^* = y/d_0$, and yawing motion, ψ, are obtained as follows:

$$\ddot{y}^* + \frac{d_{11}}{v^*}\dot{y}^* + k_{11}y^* + k_{12}\psi = 0 \tag{22}$$

$$\ddot{\psi} + \frac{d_{22}}{v^*}\dot{\psi} + k_{22}\psi + k_{21}y^* = 0, \tag{23}$$

where

$$\begin{aligned} &d_{11} = 2\kappa_y/(md_0\omega_\psi^2), d_{22} = 2\kappa_x d_0/(I\omega_\psi^2),\\ &k_{11} = (1 - l_0/l)(\omega_y/\omega_\psi)^2, k_{22} = 1,\\ &k_{12} = -2\kappa_y/(md_0\omega_\psi^2), k_{21} = 2d_0^2\kappa_x\gamma_e/(Ir_0\omega_\psi^2), \end{aligned}$$

and the dot denotes the derivative with respect to the dimensionless time, $t^* = \omega_\psi t$. v^* is the dimensionless forward speed.

8 APPENDIX B

8.1 Linear analysis at the critical speed

The characteristic equation of the governing equation is expressed as Eq. (5). At the critical speed, the characteristic roots are a pair of conjugate purely imaginary values and a pair of complex conjugates with a negative real part. Substituting $\lambda = \pm i\omega$ into Eq. (5) and separating real and imaginary parts yield

$$Real : \omega^4 - \left(\frac{d_{11}d_{22}}{v^{*2}} + k_{11} + k_{22}\right)\omega^2 + (k_{11}k_{22} - k_{12}k_{21}) = 0 \tag{24}$$

$$Imaginary : (d_{11} + d_{22})\omega^3 + (d_{11}k_{22} + k_{11}d_{22})\omega = 0. \tag{25}$$

From Eq. (25), ω is

$$\omega = \sqrt{\frac{k_{22}d_{11} + k_{11}d_{22}}{d_{11} + d_{22}}}. \tag{26}$$

8.2 Identification of creep coefficient

In order to extract the effect of the creep coefficient, we put d_{11}, d_{22}, k_{12}, and k_{21} as

$$d_{11} = d'_{11}\kappa_x, d_{22} = d'_{22}\kappa_x, k_{12} = k'_{12}\kappa_x, k_{21} = k'_{21}\kappa_x, \tag{27}$$

where d'_{11}, d'_{22}, k'_{12}, and k'_{21} are expressed based on the assumption of $\kappa_y = C_\kappa \kappa_x$ (in this study $C_\kappa = 0.8$) as follows:

$$d'_{11} = \frac{2C_\kappa}{md_0\omega_\psi^2}, d'_{22} = \frac{2d_0}{I\omega_\psi^2}, k'_{12} = \frac{2C_\kappa}{md_0\omega_\psi^2}, k'_{21} = \frac{2d_0^2\gamma_e}{Ir_0\omega_\psi^2}, \tag{28}$$

The frequency at the critical speed is independent of the creep coefficient and is expressed from Eq. (26) as

$$\omega = \sqrt{\frac{d'_{11} + k_{11}d'_{22}}{d'_{11} + d'_{22}}}. \tag{29}$$

Furthermore, from Eq. (24), we obtain the longitudinal creep coefficient as:

$$\kappa_x = \sqrt{\frac{\omega^4 - (k_{11} + k_{22})\omega^2 + k_{11}k_{22}}{\frac{d'_{11}d'_{22}\omega^2}{v_c^{*2}} + k'_{12}k'_{21}}}, \tag{30}$$

where ω is given in Eq. (29). Consequently, it is found from the above equation that the creepage is experimentally identified from the experimental critical speed and the measurable physical parameters of the wheelset system shown in section 2.

Vehicle System Dynamics Supplement 35 (2001), pp. 57- 83

High-Performance Multibody Models of Road Vehicles: Fully Symbolic Implementation and Parallel Computation

POSTIAU, T.; SASS, L.*; FISETTE, P.†; SAMIN, J.-C.

Université catholique de Louvain
CEREM: Centre for Research in Mechatronics
1348 Louvain-la-Neuve, Belgium

SUMMARY

The needs for analyzing vehicle dynamics rely on advanced modelling techniques, efficient program implementation and high-level computer environment. To match these requirements, two important assets should characterize modern modelling tools: computer efficiency (indispensable, for instance, for real time control or system optimization) and portability (indeed, multi-domain modelling is more and more required). In the frame of vehicle dynamics, a symbolic multibody approach is proposed in this paper. Vehicle model efficiency, which will be quantified through examples, results from a fully symbolic generation of the dynamical model in recursive form, and the resort to parallel computation techniques. Portability towards other computer environment is also ensured by the symbolic approach, the latter being able to export equations in a very versatile way.

1. INTRODUCTION

Multibody approach for modelling road vehicles. During the last decades, the emergence of powerful computers and reliable and user-friendly languages and sofwares led the scientific community to develop efficient computer programs to deal with vehicle dynamics. Although the domain of — road and railway — vehicles is quite specific with respect to other mechanical systems involving "interconnected bodies", one of the most popular approaches to analyse vehicle dynamical behavior is to resort to *multibody* formalisms and to the associated programs [1][2]: while being often qualified as general-purpose multibody programs, each of them has its own specificities wih respect to the *modelling* phase

*Aspirant FNRS
†Chercheur qualifié FNRS

(i.e. the generation of the geometrical, kinematic and dynamical equations) and to the *analysis* phase (i.e. a set of numerical procedure to analyse the system, such as equilibrium solution, modal analysis, time integration, system control, optimization, ...).

In the field of vehicle dynamics, a very enriching experience was proposed by the IAVSD[1] in 1991: four international benchmarks were established for railway and road vehicles. [2]. Although being of a certain level of complexity (number of degrees of freedom, 3D geometrical contact problem, stiff equations, ...) these benchmarks were still a bit of "academic" nature. Indeed, while dealing with real systems, model simplifications (to describe some compliant elements or to solve a geometrical problem,...) have been decided to make the problems compatible with computers and programs possibilities. In particular, these simplifications enabled to end up with efficient simulations, even 8 years ago [2]. However, the needs for real time simulation still exist at present for two obvious reasons:

1. The academic nature of the benchmarks does not necessarily match with industrial requirements which are presently far more demanding in terms of model refinement, in order to predict realistic behaviors (ex. the modelling of suspension dampers including valves characteristics, fluid dynamics, backlash etc... is far more realistic than a simple force/velocity damping equation [5]).

2. Standard numerical analyses like static equilibrium, modal analysis and time simulation are of course interesting but not sufficient anymore for the modern industry which, for instance, asks for vehicle optimization or active suspension control.

The two previous points are at the root of the approach pursued in this paper. Indeed, from the previous considerations, one notices the needs for *fast* and *realistic* multibody models which must be *portable* towards other scientific computer environments dealing with optimization or system control for instance.

Symbolic generation. The classical way to implement the two phases mentionned above— modelling and analysis — in a computer program is to do it in a pure numerical manner (see [1]). If this is quite obvious for the analysis phase (indeed, time integration of complex multibody systems unavoidably relies on numerical solvers !), this is not the unique possibility for the modelling phase which can be performed — totally or partially — in a symbolic way.

Symbolic multibody programs, such as ROBOTRAN [3], [4], only manipulate arithmetical operators and strings of alphanumerical characters to generate the *analytical* form of the equations in a desired syntax. From the multibody modelling point of view, these symbolic generators exhibit the same level of generality as their numerical competitors in the sense that they can handle large systems with any topology and containing up to 100 degrees of freedom.

[1] International Association for Vehicle System Dynamics

However, they are able to perform drastic simplifications, from the most trivial (addition/multiplication by zero, reduction of trigonometric expressions) to the most complex ones (elimination of superfluous equations).

In view of the needs discussed above, we claim that the symbolic approach is a valuable option: if this is quite obvious to reach better efficiency, it appears to be also the case for the *portability* point of view [6], [7]. Indeed, since a symbolic multibody program clearly separates the *modelling* phase from the *analysis* phase, it is thus straightforward to extract the symbolic model in order to export it to another environment.

Parallel computation in multibody dynamics. Various researches have been made to emphasize the benefits of parallel computation in multibody dynamics, e.g. [8], [9], [10], and in particular for the case of railway and road vehicles [11]. As clearly explained in [9], one can (or more exactly could) take advantage of parallelization in multibody dynamics at three levels, respectively named:

- *coarse grain parallelization:* at this level, large independent tasks can be identified in the model leading to "parallel threads" which can be performed on parallel computers: this is for instance the case for the subsystem segmentation technique discussed in this paper at Section 4.
- *medium grain parallelization*: at this level, smaller sets of independent computations can be isolated and performed in parallel; their origin can be the topology of the system or the nature of a specific analysis [9].
- *fine-grain vectorization*: by performing a data dependency analysis on the symbolically generated set of multibody equations, one can notice that the latter contain a very high level of parallelism at the scalar operation level [11], [12]: this is denoted as fine grain vectorization for which a specific architecture can be used to speed up the original serial computation [10].

Paper Organization. Section 2 will describe the multibody formalism underlying the symbolic program ROBOTRAN, including road vehicles specificities. Section 3 will then present the recent developments related to the fully symbolic computation and reduction of multibody models containing non linear kinematic loop constraints. Section 4 is devoted to multibody model parallelization, mainly for vehicle applications, at coarse grain level. Finally, Section 5 proposes two illustrative examples: the parallelization of the ILTIS vehicle time simulation and the fully symbolic model of a modern car equipped with multi-links suspensions.

2. MULTIBODY APPROACH

2.1. Tree-like Systems

As illustrated in Figure 1, a multibody system is considered in our formalism as a set of bodies interconnected by massless joints in which *relative motion* — described by relative coordinates q — can take place. At least one of the joints is directly connected to an inertial body called the *base* to which the inertial frame $\{\hat{\mathbf{I}}\} = \{\hat{\mathbf{I}}_1, \hat{\mathbf{I}}_2, \hat{\mathbf{I}}_3\}$ is attached. A body-fixed frame $\{\hat{\mathbf{X}}^i\}$ is rigidly attached to each body i of the system. For a tree-like structure, the notion of *body filiation* is a practical concept to set up the topology of the system, and in particular to define kinematic chains of bodies. As shown in Figure 1, this is the purpose of the inbody vector which contains the indexes of the parents of each body of the multibody system.

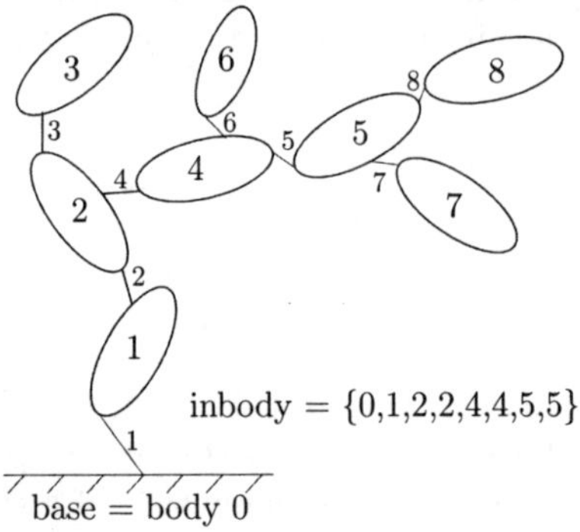

Figure 1: Tree-like multibody system: body filiation

As regards the multibody equations of motion, various well-known formalisms can be used to generate their final scalar form. Our approach is based on the Newton/Euler recursive scheme [13], [14] in which the mass matrix is recursively extracted [15]. This leads to the following well-known ODE form, for an unconstrained multibody system:

$$M(q)\ddot{q} + c(q, \dot{q}, f_{ext}, c_{ext}) = Q(q, \dot{q}) \tag{1}$$

where q denotes the generalized joint coordinates, M is the symmetric mass matrix, Q denotes the joint generalized forces and c contains the gyroscopic, centrifugal and Coriolis terms, as well as the external forces f_{ext} and torques c_{ext} (for instance, the wheel/ground contact forces of a road vehicle).
As quantatively shown in [15], [16], it is worth to note that the recursive formalism we use, while being of order $N^2 + N^3$, is still more attractive than $O(N)$ formulations for systems containing less than about 20...30 d.o.f., which is the case for our applications in vehicle dynamics[2].

[2]as soon as we resort to sub-system segmentation for large systems (see Section).

2.2. Constrained Systems

Kinematic loops. Kinematic loops are induced, for instance, by vehicle suspensions as illustrated in Figure 2. These loops induce kinematic constraints $h(q) = 0$ between the n joint variables q. In this case, using the Lagrange multiplier technique, the system of equations becomes:

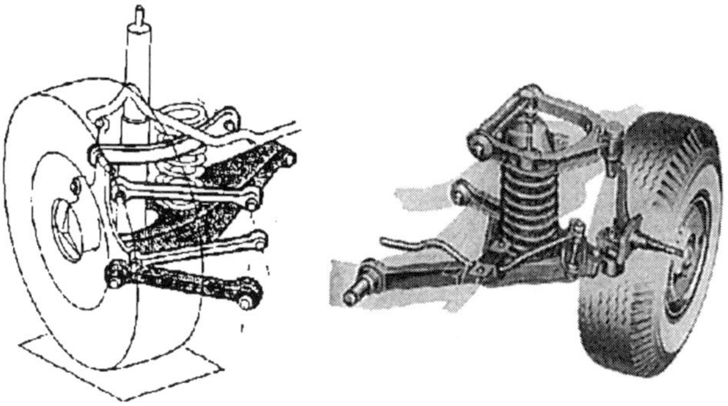

Figure 2: Kinematic loops in vehicle suspensions

$$\begin{aligned} M(q)\ddot{q} + c(q,\dot{q},f_{ext},c_{ext}) &= Q(q,\dot{q}) + J^t(q)\lambda \\ h(q) &= 0 \\ \dot{h}(q,\dot{q}) &= J(q)\dot{q} = 0 \\ \ddot{h}(q,\dot{q},\ddot{q}) &= J(q)\ddot{q} + \dot{J}\dot{q}(q,\dot{q}) = 0 \end{aligned} \tag{2}$$

where J is the Jacobian matrix associated with the constraints h and λ represents the corresponding Lagrange multipliers.

Reduction of the equations: coordinate partitioning method. In case of m independent constraints, the reduction to a purely differential form can be obtained by means of the *coordinate partitioning method* [17].

Assuming that the constraints $h(q) = 0$ are independent, the Jacobian matrix J has full rank m. In this case, m coordinates can be expressed at least *locally* as functions of the $(n - m)$ others. By this way, it becomes possible to reduce the original DAE system (2) to a set of $(n - m)$ differential equations in these $(n - m)$ independent coordinates. This reduced set will represent the equations of motion of the constrained multibody system.
Let us summarize the steps to obtain these equations.

After reordering the variables, the vector q of generalized coordinates (and the constraints Jacobian accordingly) can be partitioned as folows:

$$q = \begin{pmatrix} u \\ v \end{pmatrix} \; ; \; J = \begin{pmatrix} J_u & J_v \end{pmatrix} \tag{3}$$

where u denotes the subset of $(n-m)$ independent coordinates.
The m by m matrix J_v is regular because v has been chosen to be the subset of dependent coordinates. Let then partition the generalized mass matrix M and the vector c accordingly.

$$\begin{pmatrix} M_{uu} & M_{uv} \\ M_{vu} & M_{vv} \end{pmatrix} \begin{pmatrix} \ddot{u} \\ \ddot{v} \end{pmatrix} + \begin{pmatrix} c_u \\ c_v \end{pmatrix} = \begin{pmatrix} Q_u \\ Q_v \end{pmatrix} + \begin{pmatrix} {J_u}^t \\ {J_v}^t \end{pmatrix} \lambda$$

where ${J_v}^t$ refers to the transpose of the matrix J_v. Since this matrix is regular, one obtains by eliminating the unknowns λ:

$$\begin{aligned} & \begin{pmatrix} M_{uu} & M_{uv} \end{pmatrix} \begin{pmatrix} \ddot{u} \\ \ddot{v} \end{pmatrix} \\ + {B_{vu}}^t & \begin{pmatrix} M_{vu} & M_{vv} \end{pmatrix} \begin{pmatrix} \ddot{u} \\ \ddot{v} \end{pmatrix} + c_u + {B_{vu}}^t c_v \quad = \quad Q_u + B_{vu}^t Q_v \end{aligned} \tag{4}$$

where $B_{vu} \triangleq -(J_v)^{-1} J_u$.

Then using the first and second derivatives of the constraints, the generalized velocities and accelerations $\dot{v}$ and $\ddot{v}$ are respectively given by:

$$\dot{v} = B_{vu}\dot{u} \tag{5}$$

$$\ddot{v} = B_{vu}\ddot{u} + b \quad \text{with} \quad b \triangleq -J_v^{-1}(\dot{J}\,\dot{q}) \tag{6}$$

and can also be eliminated from the differential equations (4). This produces the *reduced* system:

$$\begin{aligned} &\left(M_{uu} + M_{uv}B_{vu} + {B_{vu}}^t M_{vu} + {B_{vu}}^t M_{vv} B_{vu}\right)\ddot{u} \\ &\qquad - \left(M_{uv} + {B_{vu}}^t M_{vv}\right) b + c_u + {B_{vu}}^t c_v = Q_u + B_{vu}^t Q_v \end{aligned}$$

which can concisely be written as:

$$\mathcal{M}(u,v)\ddot{u} + \mathcal{C}(\dot{u},u,v) = \mathcal{Q} \tag{7}$$

The algebraic constraints still have to be solved in order to eliminate the dependent variables v from (7). While analytical solutions can exist for specific cases, general kinematic loops require a numerical procedure to be solved: the Newton-Raphson iterative algorithm can be used for successive estimations of v:

$$v^{k+1} = v^k - (J_v)^{-1} h \tag{8}$$

where the right hand side is evaluated for $v = v^k$ and the values of u corresponding to the instantaneous system configuration.
Thanks to this final numerical elimination, the set of purely differential equations (7) constitutes the equations of motion of the constrained system described in terms of the $n-m$ independent generalized coordinates u. The resulting flowchart of the numerical integration scheme of constrained multibody systems is illustrated in Figure 3 in which the *model*, in the sense used in the introduction of the paper, is contained in the dashed box and will be the subject of the fully symbolic generation described in Section 3.

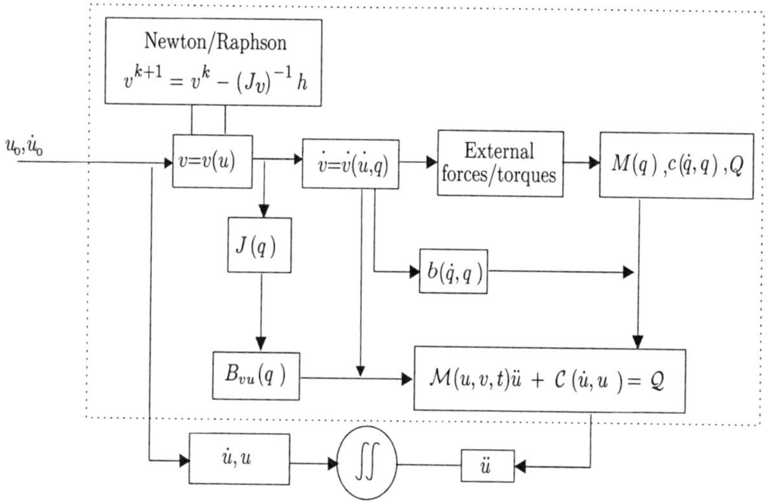

Figure 3: Time integration of reduced dynamical equations

2.3. External forces - wheel/ground model

Definitions and hypotheses. In the present multibody approach, inspired from the wheel/rail formalism proposed in [19], we assume that each wheel is a *leaf* body of the multibody structure, as shown in Figure 4.

The hypotheses underlying the proposed wheel/ground model are the following:

- The contact between the tire and the ground is assumed to be punctual and this "equivalent" point of contact Q is unique on the wheel,
- The wheel circumference has no thickness: from the geometrical point of view, the wheel (plus tire) can be assimilated to a flat disc with a variable radius,
- The normal and shear stresses taking place in the real contact patch will be replaced by a pair of equivalent force and torque vectors applied to the wheel at the contact point Q,
- The ground is assumed to be infinitively rigid and such that its profile z be known analytically in the form $z = z(x, y)$. In Figure 4, $\hat{\mathbf{n}}_g$ denotes the unit vector normal to the ground at the contact point.

To locate the geometrical contat point Q on the wheel, i.e. to find the radial vector $\mathbf{w}$ (see Figure 4), we introduce a so-called "geometrical wheel" (body frame $\{\hat{\mathbf{X}}\}$) obtained by locking the axial rotation of the real one (body frame $\{\hat{\mathbf{Y}}\}$). Two auxiliary variables are then introduced to locate, on this wheel,

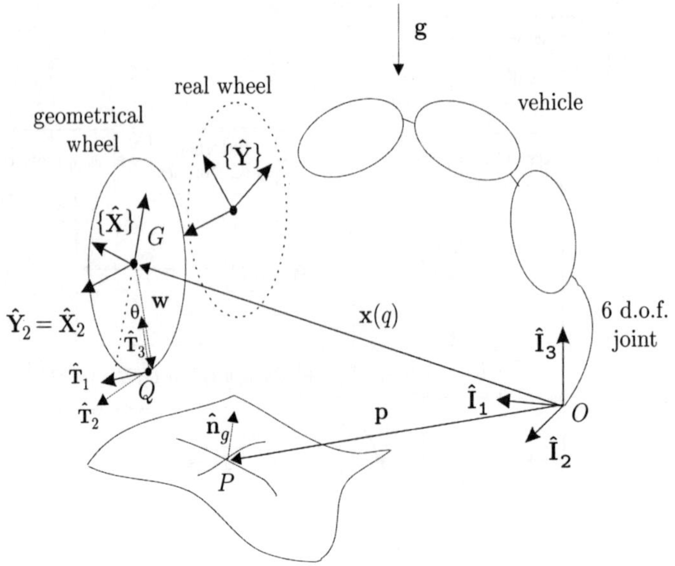

Figure 4: Wheel kinematics

the *geometrical* contact point Q in accordance with the previous hypotheses: θ which allows to locate the point on the geometrical wheel circumference and r_w which represents the (time-varying) radius of the wheel.

According to Figures 4 and 5, the wheel contact frame $\{\hat{\mathbf{T}}\}$ is introduced such that $\hat{\mathbf{T}}_1$ is tangent to both the wheel and the ground at the contact point and $\hat{\mathbf{T}}_3$ lies in the wheel plane. To be consistent with force contact models, we also use the so-called ground contact frame $\{\hat{\mathbf{R}}\}$ which is such that:

$$\hat{\mathbf{R}}_1 = \hat{\mathbf{T}}_1 \; ; \hat{\mathbf{R}}_3 = \hat{\mathbf{n}}_g \;\; ; \hat{\mathbf{R}}_2 = \hat{\mathbf{R}}_3 \times \hat{\mathbf{R}}_1$$

Contact point: geometrical solution. Although the contact point solution will not be considered as a set of kinematic constraints, the problem at the algebraic level is of the same type since it consists in solving — with respect to the auxiliary variables θ and r_w — the geometrical problem, namely: find a punctual and tangent contact between the geometrical wheel and the profiled ground. Mathematically speaking, this is ensured by two equations. Denoting $\mathbf{u}(q, \theta, r_w) = u_1\hat{\mathbf{I}}_1 + u_2\hat{\mathbf{I}}_2 + u_3\hat{\mathbf{I}}_3$ the absolute position vector of a "candidate" point of contact Q on the geometrical wheel,

1. the contact point Q belongs to the ground profile:

$$g_1(r_w, \theta) \triangleq u_3 - \mu(u_1, u_2) = 0, \tag{9}$$

2. at the contact point Q, the geometrical wheel circumference is tangent to

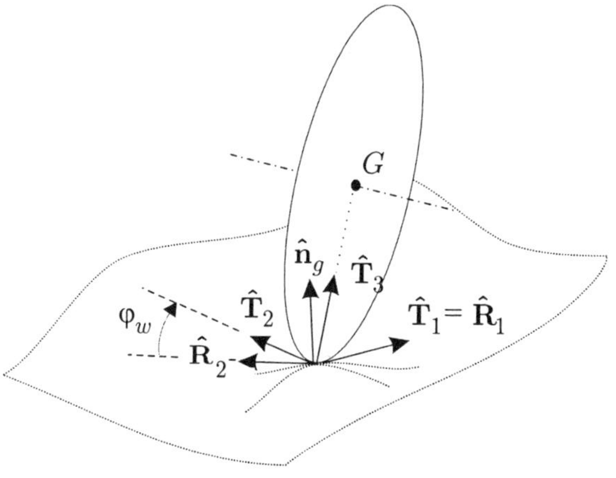

Figure 5: Wheel/ground contact frames

the ground:

$$g_2(r_w, \theta) \triangleq \hat{\mathbf{T}}_1 . \hat{\mathbf{n}}_g = 0. \tag{10}$$

Equations (9) and (10) must be solved for each wheel at any time with respect to r_w and θ. Being non linear in these variables, the latter can be found iteratively by a Newton-Raphson algorithm[3], in a way similar to equation (8) for the loop constraints.

Wheel/ground forces and torques.

Wheel ground contact kinematics. Any wheel/ground force contact model which can be used in the frame of the present approach relies on specific geometrical and kinematic quantities which must be evaluated for each wheel configuration. These quantities are:

- The *camber angle* φ_w (see Figure 5) which represents the angle between the wheel plane and the normal vector $\hat{\mathbf{n}}_g$:

$$\varphi_w = \tan^{-1}\left(\frac{\hat{\mathbf{n}}_g . \hat{\mathbf{T}}_2}{\hat{\mathbf{n}}_g . \hat{\mathbf{T}}_3}\right) \tag{11}$$

- The *sideslip angle* α_w (see Figure 6) which represents the angle – measured at the contact point in the contact plane $\{\hat{\mathbf{R}}_1, \hat{\mathbf{R}}_2\}$ – between the projection

[3] For particular cases such as wheel on flat ground, one can solve equations (9) and (10) analytically.

in this plane of the velocity vector of the *geometrical* contact point $\mathbf{v}_Q$:

$$\alpha_w = tan^{-1}\left(\frac{\mathbf{v}_Q \,.\, \hat{\mathbf{R}}_2}{\mathbf{v}_Q \,.\, \hat{\mathbf{R}}_1}\right) \tag{12}$$

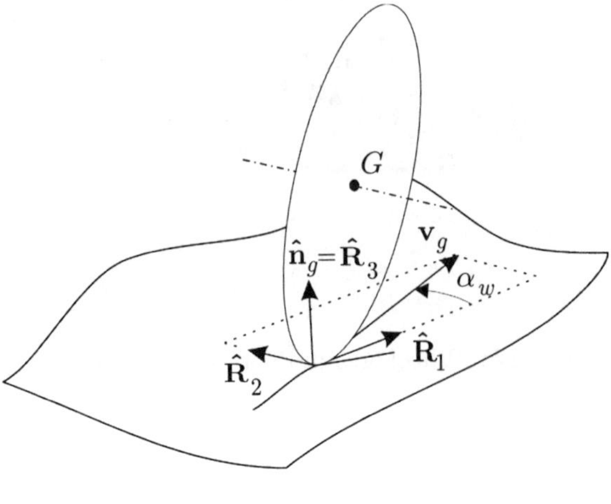

Figure 6: Wheel sideslip angle α_w

- The *longitudinal slip velocity* γ_w which is the projection on $\hat{\mathbf{R}}_1$ of the velocity of the *material* point of contact of the real wheel coincinding with $Q : \gamma_w = (\dot{\mathbf{x}} + \boldsymbol{\omega}_y \times (-\dot{r}_w \hat{\mathbf{T}}_3)) \,.\, \hat{\mathbf{R}}_1$ in which $\dot{\mathbf{x}}$ is the absolute velocity of the wheel centre, $\boldsymbol{\omega}_y$ denotes the angluar velocity of the real wheel and $-\dot{r}_w \hat{\mathbf{T}}_3$ locates the contact point with respect to the wheel centre (see Figure 4). Since $\hat{\mathbf{T}}_3 \,.\, \hat{\mathbf{R}}_1 = 0$, the latter expression reduces to:

$$\gamma_w = (\dot{\mathbf{x}} + \boldsymbol{\omega}_y \times \mathbf{w}) \,.\, \hat{\mathbf{R}}_1 \tag{13}$$

- The *tire penetration* p_w and penetration rate $\dot{p}_w$ which are measured along the normal vector $\hat{\mathbf{n}}_g$. From simple geometrical considerations, p_w can be expressed as:

$$p_w = \min\{0, \cos\varphi_w \, (r_w - r_{nom})\} \tag{14}$$

where r_{nom} is the nominal radius of the wheel (i.e. when unloaded).
The penetration p_w is thus negative when the wheel is in contact.
As regards the penetration rate $\dot{p}_w$, it corresponds to the tire radial dilatation rate and is obtained by projecting the vector $\dot{r}_w \hat{\mathbf{T}}_3$ along the normal vector $(-\hat{\mathbf{n}}_g)$:

$$\text{if } |p_w| > 0 \text{ then } \dot{p}_w = \dot{r}_w \cos\varphi_w \text{ , else } \dot{p}_w = 0 \tag{15}$$

$\dot{p}_w$ is thus negative when the tire sinks into the ground.

Contact force model. From our point of view, two criteria are relevant to classify different wheel/ground models:

1. Their *spectrum* with respect to different car behaviours (lateral stability, braking/traction, vertical vibration, cornering, ...): a lot of models are dedicated to one or more of these behaviors and it is often useless to use a full 3D model for analysing specific applications.

2. Their *level of refinement*: let us illustrate this point via the following example. To compute the lateral ground/tire force F_{lat} for instance, a "lateral model" in which this force only depends linearily on the *sideslip angle* α_w is sufficient to perform modal analyses of the vehicle around its quasi-static equilibrium[4]. On the other hand, when performing time simulations involving steering manoeuvres, saturation effects must be taken into account in the relation $F_{lat}(\alpha_w)$ for large slip. As soon as rolling motion of the vehicle becomes important, then the influence of the camber angle φ_w is not negligible and must be involved in the expression of F_{lat}, etc...
The "optimal" model is thus not necessarily the most sophisticated one, but the one whose degree of refinement is just enough to predict the envisaged car behaviour.

Outputs of wheel/ground models are forces and torques acting on the wheel at the equivalent contact point Q. To be coherent with the most standard models one can find in the literature, the latter vectors are expressed in the ground contact frame $\{\hat{\mathbf{R}}\}$:

$$\mathbf{F}_w = F_{long}\hat{\mathbf{R}}_1 + F_{lat}\hat{\mathbf{R}}_2 + F_{vert}\hat{\mathbf{R}}_3 \quad (16)$$

$$\text{and } \mathbf{M}_w = M_{roll}\hat{\mathbf{R}}_1 + M_{pitch}\hat{\mathbf{R}}_2 + M_{yaw}\hat{\mathbf{R}}_3 \quad (17)$$

in which components F_{long}, F_{lat} and F_{vert} represent the longitudinal, lateral and vertical forces and M_{roll}, M_{pitch} and M_{yaw} are respectively named the overturning, rolling resistance and aligning torques. Since it is not the purpose here to develop contact models, let us shortly illustrate one of them, proposed by Bakker et al. in [18],which is based on numerous experimental measurements on which non linear functions have been fitted. Figure 7 shows:

- on the left plot, the evolution of the longitudinal force F_{long} with respect to the longitudinal slip ratio $\gamma_w^* \triangleq \frac{\gamma_w}{\mathbf{v}_g \,.\, \hat{\mathbf{R}}_1}$ for different values of the sideslip angle α_w (expressed in degrees); the vertical load and the camber angle have been fixed as follows: $F_{vert} = 1000N$ and $\varphi_w = 0$.

- on the right plot, the evolution of the lateral force F_{lat} with respect to the sideslip angle α_w (expressed in degrees) for different values of the longitudinal slip ratio γ_w^* (with $F_{vert} = 1000N$ and $\varphi_w = 0$).

[4]Here, quasi-static equilibrium means that the vehicle is running at constant velocity V on a flat ground.

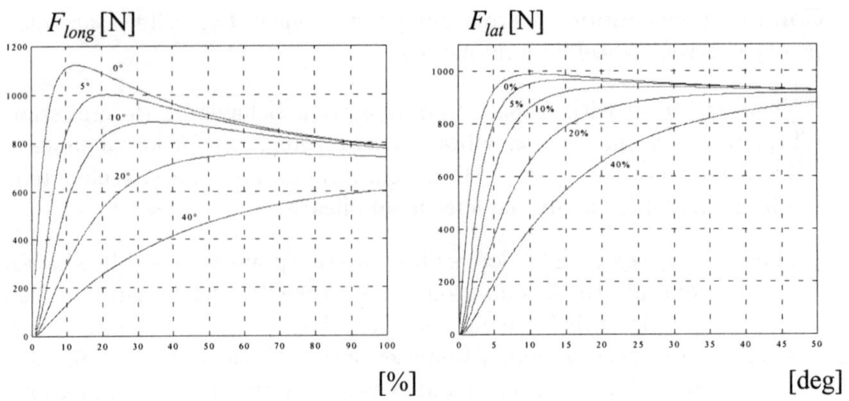

Figure 7: Wheel/ground tangent forces

On both figures, one can observe the saturation effect for large slip (all the more reduced since the slip in the other direction increases).

3. SYMBOLIC GENERATION

3.1. The ROBOTRAN program

ROBOTRAN [3], [7] is a C-language program dedicated to the *symbolic* generation of the kinematic and dynamical equations of multibody systems. It is a stand-alone program which contains its own symbolic library for manipulating, comparing and simplifying alphanumeric expressions. Among the "standard" symbolic tricks of ROBOTRAN, let us mention the elimination of redundant expressions, the recursive simplification of trigonometric terms, the use of auxiliary recursive variables, ...

The restricted field of applications of ROBOTRAN (multibody system equations) allowed us to endow it with symbolic capabilities which are not necessarily available in more general symbolic softwares[5]. For instance — and this appears to be a limitation of the latter — the memory space required by ROBOTRAN is closely controlled and minimized during the symbolic process so that the symbolic generation of very large multibody models [20] is possible within a few seconds on standard computers without any difficulties.
Another specific feature of the ROBOTRAN capabilities is related to the recursive nature of the multibody equations whose interdependency can give rise to drastic symbolic simplifications when using linked lists based programming techniques. Indeed, for most of the practical applications, about 30% of the vector recursive equations are superfluous for the final result: recursive programming techniques allow to eliminate all of them symbolically [15].

[5] with respect to symbolic commercial packages like Mapple or Mathematica

3.2. Fully symbolic generation of multibody models

Introduction. Up to now, the symbolic capabilities of the ROBOTRAN program were only exploited for generating the direct dynamics and direct kinematics of tree-like systems: this means that for closed loop systems (like vehicle suspensions), most of the "blocks" of Figure 3 were performed numerically on the basis of symbolic elements, like the mass matrix or the constraints Jacobian. While being far more efficient than a pure numerical model [16], we recently noticed that there was still a lot of superfluous operations in the numerical processes. The latter could be avoided if the symbolic recursive generation mentionned above would be applied *globally* on the model, i.e. from the input $(u, \dot{u})$ to the output $(\ddot{u})$ in case of direct dynamics; this has motivated the recent developments of ROBOTRAN described in this Section.
We will thus explain the different steps which the symbolic generation process must handle so as to obtain the full symbolic model of a multibody system containing kinematic loops, binding forces and wheels. The sequence of the symbolic computation matches with the block-diagram of Figure 3 which originally had a mixed numerical/symbolic nature. First of all, to reach this goal, expressions appearing in the block-diagram 3 (such as $J_v^{-1}.h, B_{vu}, ...$) must be computed symbolically on the basis of well-known algorithms used in algebra for solving linear equations system. Since the corresponding algorithms have a *recursive* nature, their computation (and simplification!) were easily implemented in the ROBOTRAN library. In other words, the first step of the work was to enlarge the recursive symbolic capabilities of ROBOTRAN to all these operations, previously performed numerically.

Constraints solving at position level. As explained in Section 2.2 (see (8)), the resolution of the non linear constraints at position level is done using the Newton-Rapson iterative scheme. The right-hand-side of the iterative formula $v_{k+1} = v_k - J_v^{-1}h$ must thus be generated symbolically, on the basis of:

- the constraints themselves $h(u, v) = 0$
- the constraints Jacobian J, partitioned on the basis of equation (3).

The term $J_v^{-1}h$ is obtained by symbolic resolution (with respect to x) of the linear system $J_v x = h$. Of course, the LU factorisation of the J_v part of the Jacobian matrix requires that the dependent variable v be a priori chosen and ordered according to equation (3) so that J_v be regular for any configuration[6]. The convergence of the Newton/Raphson iterative algorithm is controlled by evaluating the norm of h: this is effectively done by symbolically computing the scalar product $h^t\, h$.

As regards the loop constraints themselves, three possibilities to establish them have been systematized for convenience in ROBOTRAN [16]. They are

[6] Of course, this can be checked only numerically

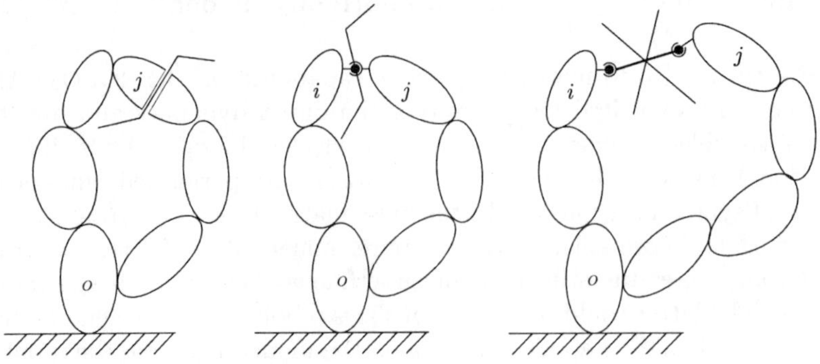

a. Cut of a body b. Cut of a spherical joint c. Cut of a connecting rod

Figure 8: Loop constraints - cutting procedures

depicted in Figure 8: in the first case, six constraints express the loop closure on a cutted body. Three constraints only are needed in the second case where a perfect spherical joint is cutted. The last case is restricted to loops closed by a connecting rod whose mass can be neglected and thus replaced by a unique "distance" constraint.

All the above computations, including the computation of the iterative formula $v_{k+1} = v_k - ...$), are generated symbolically and encapsulated in a do-while[7] instruction pair so that the numerical evaluation of this set of equations repeats until the convergence of v.

Constraint resolution at velocity and acceleration levels. According to equations (5) and (6), we must also solve the whole set of constraints at velocity and acceleration levels respectively. In particular, the coupling matrix appearing in (5), $B_{vu} \triangleq -(J_v)^{-1} J_u$, can be really time consuming for large models, if computed numerically. Indeed, J_u and J_v matrices are generally rather sparse (typically for a multi-link vehicle suspension) depending on the multibody topology and the chosen coordinate partitioning. Thus, the interest in using the symbolic resolution is particularly obvious for this part of the computation which consists in solving $(n-m)$ linear systems[8]:

$$(-J_v)B_{vu}^{(i)} = J_u^{(i)}, i = 1 : n - m$$

The dependent acceleration $\ddot{v}$ requires the computation of $b \triangleq -(J_v)^{-1} (\dot{J}\,\dot{q})$ appearing in equation (6). The elements J_v and $\dot{J}\,\dot{q}$ being available symbolically for every constraint (direct kinematics of sub-chains), vector b is also obtained by symbolically solving the linear system $-(J_v)\,b = \dot{J}\,\dot{q}$.

[7] whose syntax of course depends on the selected language (C, Matlab, ...)
[8] Superscript (i) denotes the ith column of the corresponding matrix

Link forces between bodies. When dealing with multibody systems in which some bodies are "point-to-point" connected by a spring and/or a damper and/or any other physical device, in most of the cases the constitutive equation of the corresponding force requires to compute the distance and relative velocity between the two "anchor" points: this can be consuming when using relative coordinates. As for loop constaints, the principle is to benefit from the symbolic approach to compute these forces and to project them in the adequate frames. The symbolic process thus applies to:

- the computation of all the anchor point positions and velocities
- the distance and relative velocity of each pairs
- after a user subroutine call (which provides the force constitutive equation), the projection of the forces in the correct body frames.

Wheel/ground contact forces. Starting from the multibody description, ROBOTRAN provides the symbolic expressions of the instantaneous wheel radius r_w, of the longitudinal slip velocity γ_w velocity and of the camber and side slip angles φ_w, α_w: all these ingredients are then passed to a user subroutine for computing the contact force on the basis of a given model. Finally, the expression of the equivalent forces and torques are projected into the wheel frame.

Up to now, the symbolic process only deals with flat ground, but we plan to extend it in the next future to the general case described in Section 2.3.: this amounts to perform the Newton/Raphson iterative procedure, already implemented symbolically for loop constraints, on equations (9) and (10).

Open-loop dynamics. This part of the symbolic process generates the mass matrix $M(q)$ and the dynamical vector $c(q, \dot{q}, ...)$ appearing in equation (1): this of course represents a large part of the model computation. The generation is based on the Newton-Euler recursive scheme, applied on the tree-like structure, from which the mass matrix is recursively extracted [15]. As explained at the end of Section 2.1, this formalism is one of the "best" candidates for simulation of multibody systems.

System reduction and resolution. This process is related to the last block of Figure 3 which, on the basis of the coordinate partitioning $q = (u \ v)^t$, performs the system reduction from system (2) to equation (7) and solves the latter with respect to $\ddot{u}$, the independent generalized accelerations. One must cmphasize the two following points:

- the computation of the reduced system (7), i.e. of the mass matrix $\mathcal{M}$ and of vector $\mathcal{C}$, is really time comsuming when performed numerically because of the sparsity of the involved matrices, meaning that the symbolic generation is very efficient for this block.

- For the reduced mass matrix $\mathcal{M}$ being symmetric and positive definite, the Choleski decomposition is used to factorize it; a final forward/backward substitution is then applied to obtain the accelerations $\ddot{u}$.

Symbolic model output. To clearly understand the symbolic approach underlying the above model generation — which makes the latter very compact and thus efficient — one must note that, although the construction of the model is based on distinct blocks as described here above (and according to Figure 3), the full model is perceived in ROBOTRAN as a *unique* list of serial recursive equations, whose inputs are constant data and variables $(u, \dot{u})$, and whose outputs are the independent accelerations $\ddot{u}$. The latter, once symbolically computed, automatically provoke the selection of the *usefull* equations (printed later on) and the elimination of the *superfluous* ones, whatever their origin: loop kinematics, open-loop dynamics, system reduction, ... This "cleaning" technique, in addition to the symbolic process which apply at a lower level to any individual equation, ensures that the resulting symbolic function is highly optimized.

4. COARSE GRAIN PARALLELIZATION OF SUB-MULTIBODY SYSTEMS

As illustrated in Figure 9, *sub-system segmentation* amounts to choosing a good compromise between a fully absolute or a fully relative coordinates approach. Furthermore, and this is one of the purpose of the present work, one can make the most of this segmentation approach to perform parallel computation, as it will be briefly discussed in this Section and quantified in Section 5.

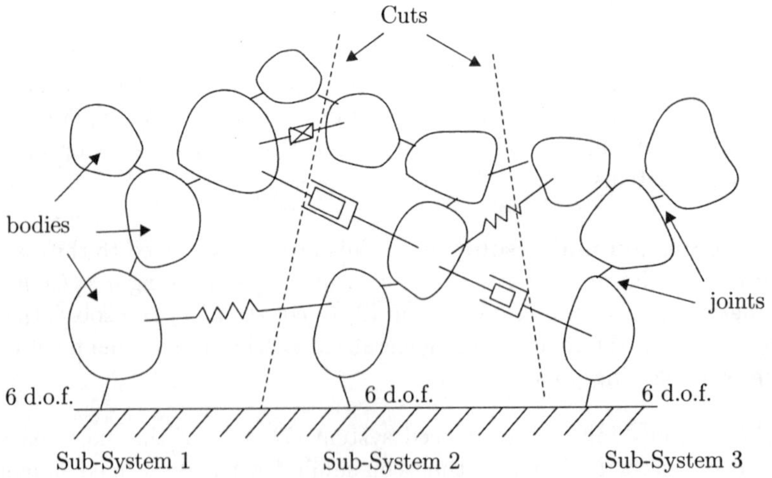

Figure 9: Segmentation into sub-multibody systems

If kinematic constraints are unavoidable to ensure the sub-system connections (at the level of the cuts *between* sub-systems, see Figure 9), a high degree of parallelism in the computations can be guaranteed if one accepts the following restriction for the coordinate partitioning:

> *Constraints between sub-systems only involve generalized coordinates belonging to the set of coordinates* $\{\{u^1\}, ..., \{u^n\}\}$ *which are — or have been selected — as independent with respect to the sub-system internal constraints.*

Indeed, as we shall see, this hypothesis allows to separate both the constraint resolution and the equation reduction into two distinct steps: one at the level of the sub-systems and the other one at the level of the global system.
If one denotes $h(u) = 0$ the set of (independent) constraints *between* the sub-systems variables $u \triangleq \{\{u^1\}, ..., \{u^n\}\}$, the reduced equations of motion (7) of sub-system i become:

$$\mathcal{M}^i \, \ddot{u}^i + \mathcal{C}^i_{red} = \mathcal{Q}^i + \left(H^i\right)^t \mu \tag{18}$$

where μ are the Lagrange multipliers associated with the constraints between sub-systems: $h(u) = 0$ and $H^i \triangleq \frac{\partial h}{\partial (u^i)^t}$ denote the columns-block of the Jacobian matrix $H = \frac{\partial h}{\partial u^t}$ related to sub-system i.
Considering the whole system, one can write:

$$\mathcal{M} \, \ddot{u} + \mathcal{C} = \mathcal{Q} + H^t \mu \tag{19}$$

with $\mathcal{M} = diag(\mathcal{M}^1, \mathcal{M}^2, ..., \mathcal{M}^n)$.

The system (19) is the result of the first reduction step, which can be parallelized sub-system per sub-system as shown in Figure 10. Since constraints $h(u)$ induce a new interdependency between the variables u, one will resort to a *second* reduction step by partitioning u into dependent variables u^D and independent variables u^I and write:

$$u^D = u^D(u^I), \; \dot{u}^D = \dot{u}^D(\dot{u}^I, u), \;\; \ddot{u}^D = \ddot{u}^D(\ddot{u}^I, \dot{u}, u) \tag{20}$$

This second reduction based on equations (20) is applied to (19) and finally leads to the following ODE system in terms of the minimal set of variables u^I :

$$\mathcal{M}^I \, \ddot{u}^I + \mathcal{C}^I = \mathcal{Q}^I \tag{21}$$

Figure 10 describes the general flow-sheet, leading to equations (21) solved with respect to $\ddot{u}^I$, which can then be time integrated. One notes that the treatment of the constraints between sub-systems — i.e. resolution (20) and reduction (21) — is performed upstream and downstream from the parallelized blocks: this guarantees a coarse grain level parallelism.

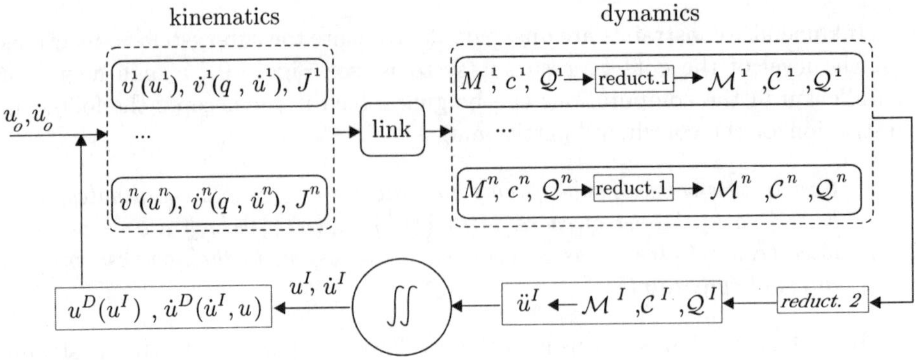

Figure 10: Parallel computation of sub-multibody systems

5. APPLICATIONS

5.1. Iltis benchmark

Figure 11: The ILTIS military vehicle

Description of the system. The ILTIS military vehicle is represented in Figure 11. It was selected as an international benchmark for multibody dynamics softwares [2]. This vehicle has a similar suspension arrangement at the front and rear. As shown in Figure 12, a leaf spring, a wheel-carrier and a $A-$arm form a four bar mechanism with the chassis. A shock absorber in parallel with a coil spring connects the $A-$arm to the chassis and, according to the benchmark, the leaf spring flexibility is taken into account by an "equivalent" vertical spring acting between the chassis and a "rigid" leaf spring. Finally, the chassis is equipped with a stiff bump stop to limit the vertical deformation of the suspension (see Figure 12). For more details, the reader can refer to [2].

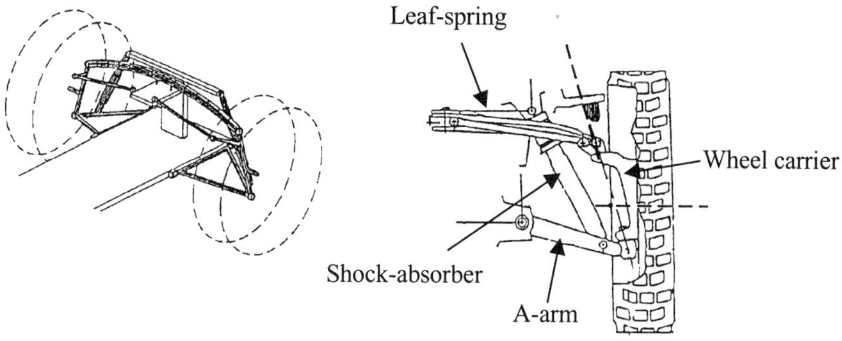

Figure 12: ILTIS vehicle suspension

Multibody model. Seen as a whole, the ILTIS jeep can be considered as a global multibody system, as depicted in Figure 13_{left}, for which a relative coordinates approach seems to be appropriate.

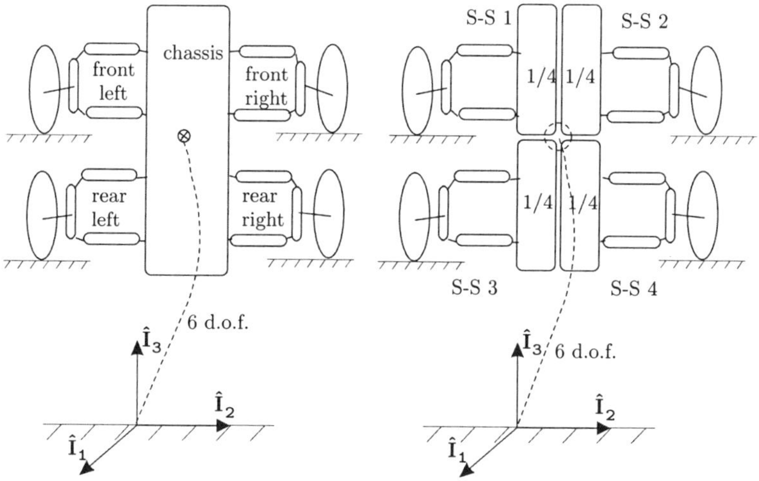

Figure 13: Modelling the ILTIS vehicle (a) global model; (b) sub-system model

However, according to Section 4, we propose to cut the chassis into 4 parts leading to 4 sub-systems which consist of $\frac{1}{4}$ chassis + 1 suspension (see Figure 13_{right}). The real model is then obtained by applying the following trivial constraints to the 4×6 d.o.f. which precede the $\frac{1}{4}$ chassis:

$$\text{For } i = 1:6,\ u_i^1 = u_i^2 = u_i^3 = u_i^4 \tag{22}$$

which represents a set of 18 independent constraints.

In this way, the model matches with the flow-sheet of Figure 10 apart from the fact that there is no need for a contributive link in the present case, and thus no synchronization point between the kinematic and dynamical blocks.

Results and Parallel computation performances The proposed experiment simply consists in performing a time integration of the jeep moving on a given profiled road: an "anti-phase cosine" bump (see [2]).

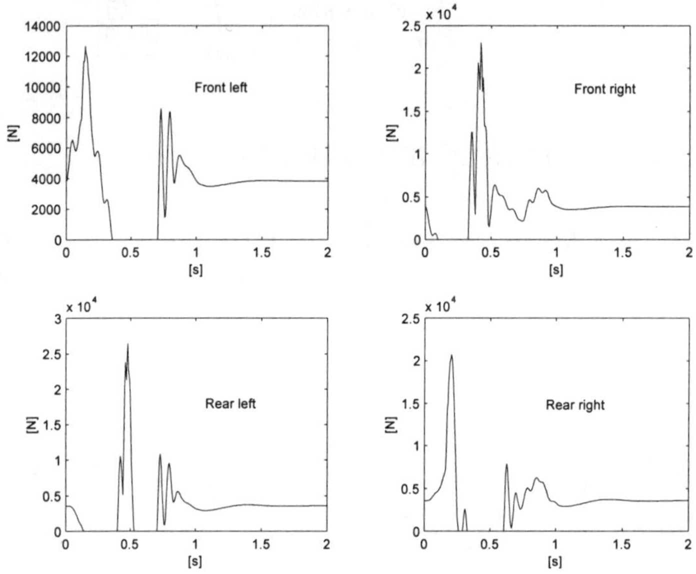

Figure 14: Wheel/ground normal forces: anti-phase cosine bump

For the time simulation at $V = 10\frac{m}{\text{sec}}$, Figure 14 shows for instance the wheel/ground normal forces: loses of contact are obvious on these figures; the high peaks in the normal force curves results from the bump stops which are made of very stiff rubber and induce high frequency transient responses when hitted by the suspension. Figure 15 illustrates an animation of the vehicle on

Figure 15: The ILTIS vehicle on the in-phase cosine bump

the in-phase cosine bump.

Coarse Grain Parallel Computation. CPU time performances are noticeably increased when running the model on a multi-processor architecture. The computer at our disposal is a CONVEX Exemplar supercomputer which uses Scalable Parallel Processing ("SPP") : each group of processors (HP PA-RISC 7100) shares a common memory of min. 256 Mbytes. In the case of a low number of processors, this architecture exhibits high synchronization and data communication performances with respect to a classical distributed memory system.
An attractive feature of this computer is the ease of program parallelization by the introduction of simple compiler directives inside the source files. In particular — *and this is essential for the sub-system segmentation approach* — directives can force the computer to parallelize tasks, without resorting to the compiler's intrinsic capabilities. These directives are quite advisible in our case since the independency of the blocks of Figure 10 is well-established and we can thus thus unequivocally allocate the corresponding tasks to distinct processors.
For instance, the FORTRAN external subroutine which computes the function (i.e. the model) to be time integrated has the following form, for the ILTIS jeep:

```
subroutine jeep_paral(time,y,yp)
...
C$DIR BEGIN_TASKS
call front_left(time,y,yp)
C$DIR NEXT_TASKS
call front_right(time,y,yp)
C$DIR NEXT_TASKS
call rear_left(time,y,yp)
C$DIR NEXT_TASKS
call rear_right(time,y,yp)
C$DIR END_TASKS
...
```

Since the 4 sub-systems are identical in size (see Figure 13_{right}), this task distribution is obviously well balanced and should normally lead to considerable CPU parallel performances improvement.

Figure 16 shows the increase in performance when using 1, 2, 3 and 4 processors successively to execute the code with the 4 previous task directives.

As a major conclusion, one notes an acceleration factor of $\frac{100}{29.3} = 3.4$ from the serial code to the 4-processor parallel code. This is a very encouraging result since one must point out that some parts of the computation are still performed sequentially, namely (see Figure 10) the second reduction step, the resolution in terms of the accelerations $\ddot{u}^I$ and of course the time discretization scheme underlying the integration routine. The absence of performance improvement between 2 and 3 processors is simply due to the fact that the above compiler directives force each of the 4 tasks to be performed by one processor: thus, in this context, passing from 2 to 3 processors cannot improve the CPU perfomance (see Figure 16).

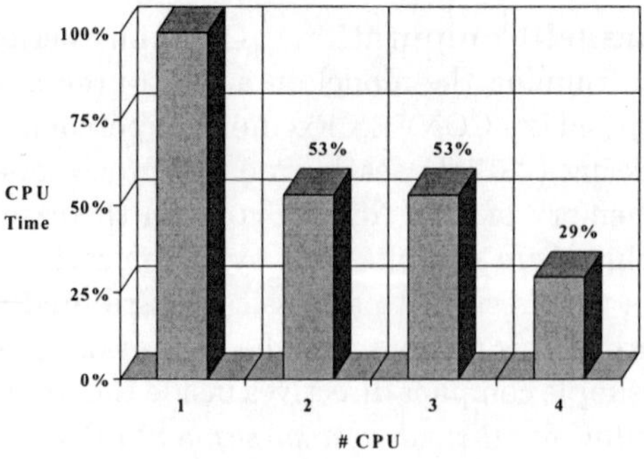

Figure 16: Coarse grain parallelism of multibody system: CPU performance

5.2. A modern car: real time simulation

Multibody model. This second example deals with the modelling of a full modern car equipped with multi-link suspensions at front and at rear. Figure 17

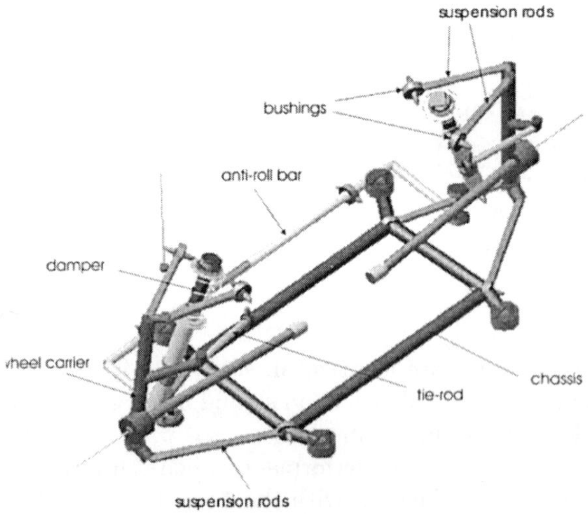

Figure 17: Front wheel-axle unit

shows the front wheel-axle unit consisting of a part of the chassis, the suspension rods, the dampers, the anti-roll bar, the tie rod and the rubber bushings at the various connection. From the multibody modelling point of view, this vehicle is rather more complex than the previous one since it involves:

- more than 80 generalized coordinates,

- multi-link suspensions (rear and front), which induce 16 three-dimensional kinematic loops,
- compliant bushing elements[9] between the various rods and the chassis, the damper and the wheel carrier: these bushing induce internal compliant torques which must be characterized in terms of their deformations and deformation rates.
- a longitudinal/lateral wheel/ground model with saturation effect.

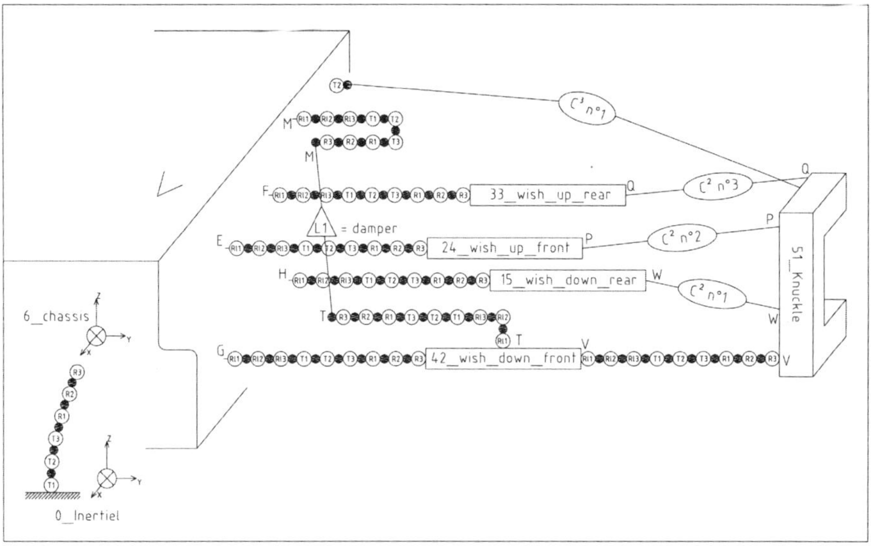

Figure 18: ROBOTRAN multibody model of a multi-links suspension

Figure 18 illustrates the multibody representation of one suspension (front left) in which each circle denotes a joint involving a generalized coordinate.

Illustrative results. Two numerical results are shown in figures 19_a and 19_b. For obtaining Figure 19_a, a static parametrization was performed which computes the carbody roll when subjected to an increasing centrifugal effect (quantified in the plot by an equivalent lateral force [N] acting at the vehicle centre of mass). The study was done for three situations: the car without anti-roll bars, with the standard anti-roll bar and with a ten-time stiffer one. The interest in this first static simulation is that each point of the curves only requires the computation of the vehicle equilibrium with a Newton/Raphson algorithm.

[9] along rotational degrees of freedom

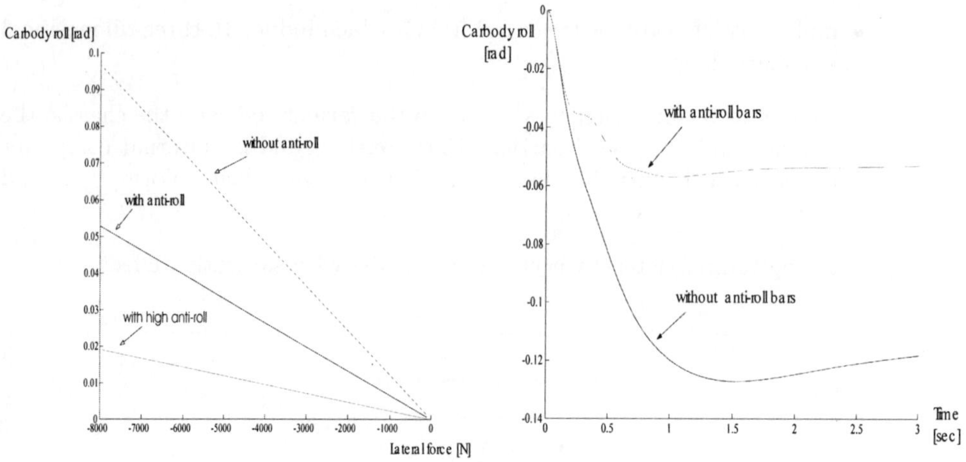

Figure 19: Simulation of a modern car - a: quasi-static equilibrium - b: entry curving

The second Figure 19_b is a time simulation in entry curving (forward velocity $V = 20\frac{m}{\text{sec}}$, average curve radius $= 50\,m$). After a transient response, the vehicle reachs quasi-static equilibrium which reasonably corresponds to its "quasi-static" equivalent in Figure 19_a.

Computer performances. To illustrate both the efficiency and the user-friendliness of the fully symbolic generation of car models detailed in Section 3.2, a double line-change simulation of the car (without anti-roll bar) is performed over 10 s. in the SIMULINK environment.For this purpose, the full car model is generated symbolically by ROBOTRAN as a single C function[10], which can easily be introduced in a S-function template in C syntax. Figure 20 illustrates the SIMULINK window related to the car simulation. Ten seconds simulated required approximately 11 sec on a Pentium3 - 450 Mhz computer. One can thus note that nearly real time simulation (=*efficiency*) is obtained in SIMULINK (=*userfriendliness*) with a minimum of effort to introduce the model, thanks to its fully symbolic generation.

6. CONCLUSION

The contribution of this paper in the field of vehicle dynamics is to show that multibody symbolic models and parallel computation techniques are very efficient tools for performing time simulation of complex vehicles. This efficiency is mandatory for present industrial researches in dynamics, involving for instance hybrid modelling or system control and optimzation.

[10] An additional function devoted to the wheel/ground specific model is of course supplied and called within the symbolic model.

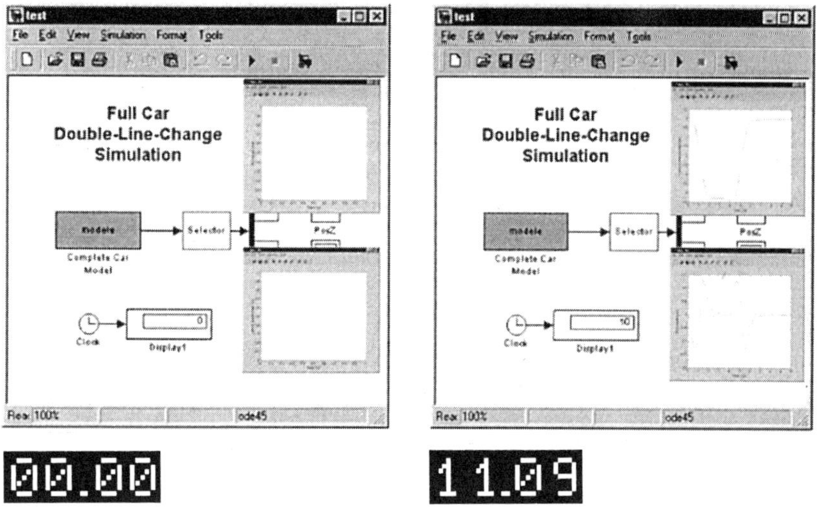

Figure 20: SIMULINK time simulation of a double line-change

In terms of parallel computing, it has been demonstrated that "large" multibody systems are good candidates for coarse grain level parallelization by judiciously splitting them into sub-systems. This operation enables to isolate large independent blocks in the simulation program. The — a priori — ensurance that these blocks are fully independent from one another is a major advantage since this allows the user to be sole manager of the way the program will be distributed among several processors. If his choice of segmentation leads to well balanced blocks, he can expect better results than by intrusting this distribution task to the compiler, as generally done. This has been validated and quantified via the simulation of a full vehicle on a parallel computer.

As regards the symbolic approach itself, the efficiency of our previous models have been greatly enhanced thanks to the global symbolic treatment of the full model, i.e. including loop closure, wheel kinematics, multibody dynamics and equation reduction. Rather than splitting these steps into distinct blocks which are thus only locally optimized symbolically, the symbolic elimination is applied recursively on the *global* set of equations and enables to produce a very compact multibody model.

ACKNOWLEDGEMENTS

This research has been sponsored by the Belgian Program on Interuniversity Attraction Poles initiated by the Belgian State - Prime Minister's Office - Science Policy Programme (IUAP -24). The scientific responsability is assumes by its authors.

REFERENCES

[1] Schiehlen, W., Multibody System Handbook, W. Schiehlen, ed., Berlin, Springer-Verlag, 1989

[2] Kortüm, W., Sharp, R.S., Multibody Computer Codes in Vehicle System Dynamics, Supplement to Vehicle System Dynamics, Vol. 22, Swets and Zeitlinger, Amsterdam, 1993

[3] Maes, P., Samin, J.C., Willems, P.Y., Robotran, Multibody System Handbook, W. Schiehlen, ed., Berlin, Springer-Verlag, 1989, pp. 225-245

[4] Schiehlen, W., Advanced Multibody System Dynamics: Simulation and Sofware Tools, W. Schiehlen, ed., Kluwer Academic Publishers, Dordrecht, 1993

[5] Duym, S., Stiens, G. and Reykbrouck, K., Evaluation of Shock Absorber Models, Vehicle System Dynamics 27(2), 1997, pp 109-127

[6] Fisette, P., Samin, J.C., MBSOFT: a symbolic/numerical multibody program for analysing mechanical and mechatronic systems, ESM'98, Manchester, june 16-19, 1998

[7] Fisette, P., Lipinski, K. and Samin, J.C., Symbolic Modelling for the Simulation, Control and Optimization of Multibody Systems, in Advances in Multibody Systems and Mechatronics, Kecskeméthy, A. et al. (eds), Institüt für Mechanik und Getriebelehre Technische Universität Graz, Graz (1999), pp 139-174

[8] BAE, D.S., Haug, E.J. and Kuhl, A., A recursive formulation for constrained mechanical systems: Part 3, Parallel Processor Implementation, Mechanics of Structures and Machines 16(2), 1988, 249-269

[9] Eichberger, A., Führer, C. and Schwertassek, R., Recent advances in the numerical integration of multibody systems, in: Schiehlen, W., Advanced Multibody System Dynamics: Simulation and Software Tools, W. Schiehlen, ed., Kluwer Academic Publishers, Dordrecht, 1993

[10] Kircanski, N., Petrovic, M. and Vukobratovic, M., Parallel computation of symbolic robot models and control laws: theory and application to transputer networks, J. Robotics Syst. 10(3), 1993, 345-368

[11] Fisette, P., Péterkenne, J.M., Contribution to Parallel and Vector Computation in Multibody Dynamics, Parallel Computing, vol. 24 (1998), pp 717-728

[12] Postiau, T., Fisette, P. et Legat, J.D., Fine Grain Parallelization of Multibody System Equations of Motion, ParCo'99 (Parallel Computing), Delft (The Netherlands), Août 1999, pp 193-200.

[13] Luh, J.Y.S., Walker, N.W., Paul, R.P.C., On-line computational scheme for mechanical manipulators, J. Dynamics Systems, Meas. and Control, 102, pp. 69-76, 1980

[14] Renaud, M., Quasi-minimal computation of the dynamic model of a robot manipulator utilizing the Newton/Euler Formalism and the notion of augmented body, Proceedings of the IEEE International Conference on Robotics and Automation, pp. 1677-1682, Raleigh, North-Carolina, 1987

[15] Fisette, P., Samin J.C., Symbolic Generation of Large Multibody System Dynamic Equations using a New Semi-Explicit Newton/Euler Recursive Scheme, Archive of Applied Mechanics, 66 (1996), pp. 187-199

[16] Fisette, P, Génération Symbolique des Equations du Mouvement de Systèmes Multicorps et Application dans le Domaine Ferroviaire, PhD Thesis, Louvain-la-Neuve, Belgique, Mars 1994

[17] Wehage, R., Haug, E.J., Generalized Coordinate Partitioning for Dimension Reduction Analysis of Constrained Dynamic Systems, J. Mech. Design, 134, pp. 247-255, 1982

[18] Bakker, E., Nyborg, L. and Pacejka, H.B., Tyre Modelling for Use in Vehicle Dynamics Studies, SAE Technical Paper n° 870421, pp.1-15, 1987

[19] Fisette, P., Samin, J.C., A new wheel/rail contact model for independent wheels, Archive of Applied Mech., 64, pp. 180-19, 1994

[20] Fisette, P., Samin, J.C. and Valembois, R.E., Symbolic Generation of the Kinematic and Dynamic Equations of Vehicle Articulated Structures, European Journal of Mechanical Engineering., 40 n° 1 (1994), pp 9-17

[21] Kalker, J.J., Three dimensional elastic bodies in rolling contact, Dordrecht: Kluwer Academic Publishers, 1990 System Dynamics, Vol. 22, Swets and Zeitlinger, Amsterdam, 1993

Vehicle System Dynamics Supplement 35 (2001), pp. 85-100

Different actuation strategies for a yaw moment control of a passenger car

Heinzl Ph., Plöchl M., Lugner P.

Institute of Mechanics, Vienna University of Technology
Austria

SUMMARY

For a yaw moment stabilization of a passenger car in a critical driving situation the required external yaw moment can be provided by additional steering of the front or rear wheels or by unilateral braking. The numerical simulations are based on a complex vehicle model built up with a multi-body system program. For the design of the controller a simplified 4-wheel vehicle model and for the required state observer a 2-wheel vehicle model are used to derive the required stabilizing yaw moment which is then applied to the vehicle by a transformer to provide the individual inputs for each wheel, in particular additional steering angles for the front or rear wheels or unilateral braking forces. A comparison of these three transformer strategies clearly reveals their advantages and limitations.

1 INTRODUCTION

In modern high quality passenger cars the driver is supported by active control systems in critical driving situations. The required stabilizing yaw moment can be provided by additional steering, e.g. [1], at the front or rear wheels or by unilateral braking like the ESP-system by Daimler Benz or an electronic brake management, [2], or even combinations of some of these actuation strategies, e.g. [3]. Since the evaluation of the different control designs and strategies is too complex to be done by trial and error, the simulation of the mechatronic system car-controller is the only way for useful results.

Now the question arises which of the strategies is better - or more precisely - for which situation additional steering or unilateral braking provide better path deviation compensation and which are the essential effects and consequences for the vehicle behaviour. In addition, the interaction of the yaw moment control with the ABS-system of the car will be of special interest.

Consequently for the simulation a complex nonlinear vehicle and tyre model has to be used to be able to analyze the system behaviour for critical conditions where a yaw moment stabilization by adequate control activities becomes necessary. For the sake of comparison of the different actuator strategies the standard open-loop manoeuvre 'braking during cornering' is chosen as it addresses the characteristics of the control strategies in interaction with the ABS in particular.

2 VEHICLE MODEL

Due to the nowadays available multi-body system programs the model of the car including steering system and drive train is built up with such a software package (ADAMS/Car [4], [5]). This model, a scheme of which is shown in Fig.1, includes all force- and kinematic nonlinearities of the suspensions and the steering system including their individual connections to the car body that are represented by flexible rubber mounts (by stiffness and damping) or rigrid joints (by kinematic constraints).

The drive train includes quasi-static engine characteristics, clutch, gearbox and differential gears. The brake system not only provides the brake force distribution front-rear but also includes an ABS-system with individual control of each wheel, see section 2.1. For the tyre characteristics the approximation presented in [6] was introduced as a user-defined force to the ADAMS vehicle model. The transient tyre behaviour was modelled by first-order differential equations for the longitudinal and lateral direction. Since these tyre characteristics comprise all values of lateral and longitudinal slip no restrictions with respect to sliding, locked wheels or skidding of the whole vehicle are imposed.

A steering angle and brake/acceleration pedal input can be applied to the vehicle to achieve a desired initial state (e.g. a steady-state cornering) or driving condition but no driver model is involved. This simplification is sufficient for the considered open loop manoeuvres and the comparison of the different control strategies.

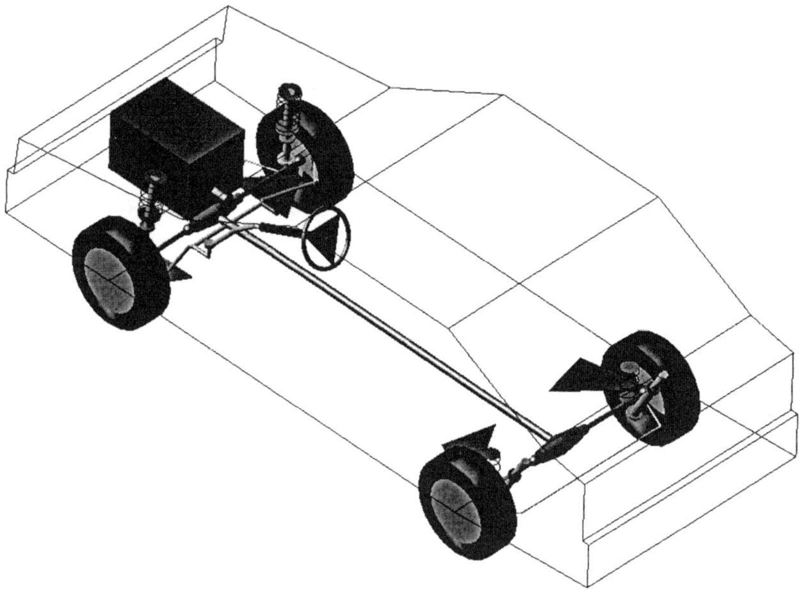

Fig. 1: Scheme of the vehicle model.

2.1 ABS and brake pressure modification

The ABS - anti lock system - is always activated to avoid the locking of the wheels. Thereby each wheel is considered separately.

The braking moment M_{Bi} at the wheel i is considered proportional to the hydraulic brake pressure p_i

$$M_{Bi} = c_{pF,R} \cdot p_i \tag{1}$$

with $c_{pF,R}$ being characteristic parameters different for front and rear wheels. When the driver induces a basic pressure $p_D = p_i$ due to a brake pedal activation the brake balance front-rear is also determined by (1).

For the measure S_i used to operate the ABS a combination of longitudinal slip s_{Li} and wheel spin acceleration $\dot{\omega}_i$ is used according to [7]

$$S_i = k_R \cdot z_{Ri} + k_s \cdot s_{Li} \,, \tag{2}$$

$$z_{Ri} = \frac{R\dot{\omega}_i}{g} \,, \quad s_{Li} = \frac{v_{ref} - \omega_i R}{v_{ref}} \tag{3}$$

$$v_{ref} = \max(\omega_i R) \,, \quad i = 1..4 \tag{4}$$

with the rolling radius R of the wheels. The constants k_R and k_s are chosen appropriately ($k_R \sim 0.2$, $k_s \sim 8$). It is assumed that the wheel spin velocity

ω_i is measured and $\dot{\omega}_i$ is determined by the time history $\omega_i(t)$. Target pressure, if not too large, is always p_D. Now, the ABS works in a way that the actual pressure p_i is to be reduced if $S_i > S_o$, not changed for $S_{u1} < S_i < S_o$ and increased for $S_i < S_{u2} < S_{u1}$. In the range of $S_{u1} < S_i < S_{u2}$ there is no specific change of ABS-activity to avoid chattering. Values used in the presented investigation are $S_0 = 1.2$, $S_{u1} = 1.0$, $S_{u2} = 0.93$. For the pressure increase $\dot{p} = 6 \cdot 10^7$ Pa/s, for the decrease $\dot{p} = -1 \cdot 10^8$ Pa/s are used, similar values to [9].

3 CONTROL SYSTEMS

Fig.2 gives an overview of the signal flow between the vehicle and the control elements. The ABS which is always activated, supplies the control circuit with the measured brake pressures and the angular velocities/accelerations of the wheels and is incorporated into the 'Vehicle' block. The 'Observer' is designed to get an estimate for the side slip angle β of the vehicle and additionally calculates approximations of the tyre forces. The 'Controller' provides the necessary correcting yaw moment to stabilize the vehicle. The controlled variable is the yaw velocity of the vehicle which should be kept equal to the steady-state yaw velocity of the linearized standard 2-wheel vehicle model. As already mentioned, the possibility to choose between unilateral braking or additional steering of the front or rear wheels demands a 'Transformer' which calculates the respective actuator inputs for the separate wheels.

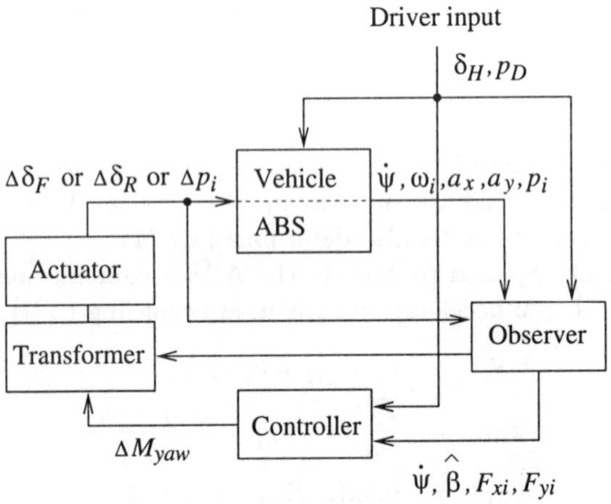

Fig. 2: Block diagram of the system configuration.

For the controller and transformer design a simplified 4-wheel vehicle model is considered, see Fig.3. Since it is not possible to measure the actual tyre forces a bilinear tyre model is used to describe the mutual influence of lateral and longitudinal tyre forces, [10], that is essential especially for braking during cornering:

$$F_{xi} = s_{Li}(k_L - \bar{k}_L|\alpha_i|)F_{zi} \quad (5)$$

$$F_{yi} = \alpha_i(k_{\alpha_i} - \bar{k}_\alpha|s_{Li}|)F_{zi} \quad (6)$$

$$\text{with} \quad \alpha_i = \delta_i - \beta - \frac{l_i \cdot \dot{\psi}}{v_{ref}}, \quad i = 1..4. \quad (7)$$

The parameters k_L, $\bar{k}_L$, $\bar{k}_\alpha$ are constant for all wheels. In order to take into account an understeering characteristic of the vehicle in a simple way, different values $k_{\alpha 1} = k_{\alpha 2} = k_{\alpha F}$ and $k_{\alpha 3} = k_{\alpha 4} = k_{\alpha R}$ for the front and the rear wheels are introduced respectively. The constants are derived from the tyre characteristics to achieve reasonable conformity with the ADAMS vehicle model. Obviously the knowledge of the normal tyre forces F_{zi}, the longitudinal slips s_{Li} (that also influence the lateral tyre forces F_{yi}) and the slip angles α_i are necessary for this kind of tyre force approximation.

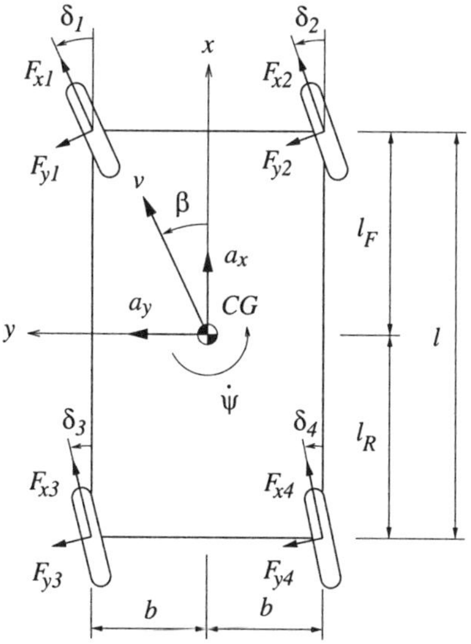

Fig. 3: Top view of the simplified 4-wheel steered vehicle model.

With the constant parameter h representing the static height of the CG above the ground, the measured longitudinal and lateral accelerations a_x , a_y and assumed equal rolling suspension stiffnesses front and rear, the normal tyre force e.g. for the front wheel 1 is estimated by

$$F_{z1} = \frac{1}{2}\left(mg\frac{l_R}{l} - ma_x\frac{h}{l} - ma_y\frac{h}{2b}\right). \tag{8}$$

Using the available measurements of $\dot{\omega}_i$, p_i from the ABS and the moment of inertia I_R of the wheel the longitudinal tyre forces F_{xi} are estimated, see [8]:

$$F_{Li} = \frac{c_{pF,R} \cdot p_i}{R} + \frac{I_R}{R}\dot{\omega}_i \tag{9}$$

$$\dot{F}_{xi} = \frac{1}{T}(F_{Li} - F_{xi}). \tag{10}$$

Equation (10) with a time constant $T = 10^{-3}s$ is introduced as a filter to avoid high frequency effects of the wheel spin changes $\dot{\omega}_i$. Now the longitudinal slip s_{Li} which is required for the lateral tyre force F_{yi} in (6) can be determined using (5)

$$s_{Li} = \frac{F_{xi}}{F_{zi}} \cdot \frac{1}{k_L - \bar{k}_L|\alpha_i|}. \tag{11}$$

In contrast to (3), where at least one slip value is 0, the calculation of s_{Li} with equation (11) is used for the yaw control. The slip angles α_i are determined by equation (7) using the measured yaw velocity $\dot{\psi}$ and reference velocity v_{ref} and an estimated side slip angle $\hat{\beta}$ of the vehicle for which an observer becomes necessary.

3.1 Observer

Since it is difficult to measure the side slip angle β of the vehicle an observer is used to find estimates $\hat{\beta}$. For the observer design the model shown in Fig.3 is simplified to a 2-wheel vehicle model by combining the left and right wheels to one front and one rear wheel in the longitudinal plane and thereby assuming

$$\delta_F = \frac{\delta_1 + \delta_2}{2}, \quad \delta_R = \frac{\delta_3 + \delta_4}{2}$$
$$F_{zF} = F_{z1} + F_{z2}, \quad F_{zR} = F_{z3} + F_{z4}. \tag{12}$$

Keeping the bilinear description for the lateral tyre forces the differential equation for the observer vehicle model is

$$\underline{\dot{x}} = \underline{A}\,\underline{x} + \underline{r} + \underline{g}_F\delta_F + \underline{g}_R\delta_R, \quad \underline{x} = \begin{bmatrix} \beta \\ \dot{\psi} \end{bmatrix}$$

$$\underline{A} = \begin{bmatrix} \dfrac{k_{\alpha F}F_{zF} + k_{\alpha R}F_{zR}}{mv_{ref}} & \dfrac{k_{\alpha R}l_R F_{zR} - k_{\alpha F}l_F F_{zF}}{mv_{ref}^2} - 1 \\ \dfrac{k_{\alpha R}l_R F_{zR} - k_{\alpha F}l_F F_{zF}}{I_z} & \dfrac{k_{\alpha F}l_F^2 F_{zF} + k_{\alpha R}l_R^2 F_{zR}}{I_z v_{ref}} \end{bmatrix}$$

$$\underline{g}_F = k_{\alpha F} \begin{bmatrix} \dfrac{F_{zF}}{mv_{ref}} \\ \dfrac{l_F F_{zF}}{I_z} \end{bmatrix}, \quad \underline{g}_R = k_{\alpha R} \begin{bmatrix} \dfrac{F_{zR}}{mv_{ref}} \\ -\dfrac{l_R F_{zR}}{I_z} \end{bmatrix}$$

$$\underline{r} = \bar{k}_\alpha \begin{bmatrix} -\dfrac{\alpha_F|s_{LF}|F_{zF} + \alpha_R|s_{LR}|F_{zR}}{mv_{ref}} \\ \dfrac{l_R\alpha_R|s_{LR}|F_{zR} - l_F\alpha_F|s_{LF}|F_{zF}}{I_z} \end{bmatrix}. \tag{13}$$

The vector $\underline{r}$ represents the nonlinearities due to the bilinear tyre model and its characteristic during braking and cornering.
With the measurement vector

$$\underline{m} \;=\; \begin{bmatrix} \dot{\beta} \\ \dot{\psi} \end{bmatrix} \approx \begin{bmatrix} \dfrac{a_y}{v_{ref}} - \dot{\psi} \\ \dot{\psi} \end{bmatrix} \tag{14}$$

and the estimated measurement vector

$$\begin{aligned} \hat{\underline{m}} \;&=\; \underline{C}\,\hat{\underline{x}} + \underline{D}(\underline{r} + \underline{g}_F\delta_F + \underline{g}_R\delta_R), \\ \underline{C} &= \begin{bmatrix} A_{11} & A_{12} \\ 0 & 1 \end{bmatrix} \;, \quad \underline{D} = \begin{bmatrix} 1 & 0 \\ 0 & 0 \end{bmatrix} \end{aligned} \tag{15}$$

the governing equation for the observer is, see [11]:

$$\dot{\hat{\underline{x}}} = \underline{A}\,\hat{\underline{x}} + \left(\underline{r} + \underline{g}_F\delta_F + \underline{g}_R\delta_R\right) + \underline{L}\,(\underline{m} - \hat{\underline{m}})\;. \tag{16}$$

The observer gain matrix $\underline{L}$ is determined using the pole assignment method and guarantees a sufficiently quick decline of the estimation error. In the control loop the estimated side slip angle $\hat{\beta}$ of the vehicle and the measured yaw velocity $\dot{\psi}$ are used.

3.2 Controller

The reference value $\dot{\psi}_d$ for the controlled yaw velocity $\dot{\psi}$ which determines the required stabilizing yaw moment is set to the steady-state value of the linear part of the 2-wheel model in (12), (13) for the actual speed and hand-wheel steering angle δ_H, [8]:

$$\dot{\psi}_d \;=\; \frac{v_{ref}\cdot\delta_H}{i_l l\,(\,1 + \frac{v_{ref}^2}{v_{ch}^2}\,)} \tag{17}$$

$$\text{with}\quad v_{ch}^2 \;=\; \frac{k_{\alpha F}k_{\alpha R}F_{zF}F_{zR}l^2}{m(F_{zR}l_Rk_{\alpha R} - F_{zF}l_Fk_{\alpha F})}$$

$$F_{zF} \;=\; mg\frac{l_R}{l}\,, \quad F_{zR} = mg\frac{l_F}{l}\,.$$

The parameter i_l represents the steering transmission ratio.

For the control design the sliding mode approach is used, [12]. The differential equation for the yaw motion is approximated by

$$\ddot{\psi} \approx \frac{M_{yaw}}{I_z} + \frac{1}{I_z}\Delta M_{yaw} \tag{18}$$

with the yaw moment ΔM_{yaw} which is to be applied additionaly. The yaw moment M_{yaw} due to the estimated tyre forces using (5), (6) and considering the 4-wheel vehicle model of Fig.3 (assuming small angles) is:

$$M_{yaw} = \sum_{i=1}^{4}[(F_{xi} - F_{yi} \cdot \delta_i)b_i + (F_{xi} \cdot \delta_i + F_{yi})l_i] \tag{19}$$

$$\text{with:} \quad \begin{array}{ll} b_i = b \cdot (-1)^i & (i = 1..4) \\ l_i = l_F & (i = 1, 2) \\ l_i = -l_R & (i = 3, 4)\,. \end{array}$$

Corresponding to (18) the actual yaw acceleration of the ADAMS vehicle is written as

$$\ddot{\psi} = f + bu \tag{20}$$

with the nonlinear function f, the control gain b and the control input $u = \Delta M_{yaw}$. Although the function f and the gain b are not known exactly, $\hat{f} = M_{yaw}/I_z$ and $\hat{b} = 1/I_z$ can be used instead, and bounds $|\hat{f} - f| \leq F$, $B^{-1} \leq \hat{b}/b \leq B$ can be specified.

In order to have a system track $\dot{\psi} \equiv \dot{\psi}_d$, a sliding surface $s = 0$ is defined by

$$s \;=\; \dot{\psi} - \dot{\psi}_d \tag{21}$$

and consequently

$$\dot{s} \;=\; \ddot{\psi} - \ddot{\psi}_d \;=\; f + bu - \ddot{\psi}_d\,. \tag{22}$$

The best approximation $\hat{u}$ of the continuous law that would achieve $\dot{s} = 0$ is

$$\hat{u} = \frac{1}{\hat{b}}\left(-\hat{f} + \ddot{\psi}_d\right). \tag{23}$$

In order to satisfy the sliding condition

$$\frac{1}{2}\frac{ds^2}{dt} = \dot{s}s \leq -\eta|s|\,, \qquad \eta = \text{constant} > 0 \tag{24}$$

despite uncertainties on the dynamics f and on the control gain b a term is added to (23), discontinuous across the surface (21):

$$u = \hat{u} - k\,\text{sgn}(s)\,. \tag{25}$$

By choosing k as

$$k \geq B(F+\eta) + (B-1)|\hat{u}| \tag{26}$$

it is guaranteed that the sliding condition (24) is verified, [12].
To avoid 'control chattering' the sgn-function in (25) is replaced by a saturation function f_{sat} with a boundary layer thickness ϕ as proposed in [12]:

$$u = \hat{u} - \hat{b}^{-1}k \cdot f_{sat}\left(\frac{s}{\phi}\right) = \Delta M_{yaw} \,. \tag{27}$$

The controller is activated as soon as a 5 % deviation of the yaw velocity $\dot{\psi}$ or of the lateral acceleration a_y from their respective reference steady-state value $\dot{\psi}_d$ or $\dot{\psi}_d \cdot v_{ref}$ is exceeded.

3.3 Transformer

The output ΔM_{yaw} of the controller can be transferred to the vehicle by a change of the tyre forces F_{xi} or F_{yi}, or the longitudinal slip or the steering angles respectively. It is supposed that the required corrections are equal for 2 wheels:

$$\left.\begin{aligned} \Delta s_{L1} &= \Delta s_{L3} = \Delta s_{Ll} \\ \Delta s_{L2} &= \Delta s_{L4} = \Delta s_{Lr} \end{aligned}\right\} \tag{28}$$

$$\Delta\delta_1 = \Delta\delta_2 = \Delta\delta_F \tag{29}$$

$$\Delta\delta_3 = \Delta\delta_4 = \Delta\delta_R \tag{30}$$

When unilateral braking is considered both equations in (28) are necessary to provide $\Delta M_{yaw} \gtrless 0$. With additional steering this can be achieved with the front wheels or rear wheels separately. These three different strategies are compared later in the results.

The use of both wheels of one side in (28) instead of the inner rear wheel or the outer front wheel only, as proposed in [13], is based on the assumption that this actuation may be advantageous with respect to the ABS.

By partial derivation of (19) with respect to the longitudinal slip s_{Li} or the steering angles δ_F or δ_R respectively the necessary equations to determine the input values for the relevant actuator can be derived. As an example, braking on the left side is necessary for $\Delta M_{yaw} > 0$:

$$\begin{aligned} \Delta M_{yaw} = \Delta s_{Ll} \sum_{i=1,3} \Bigg[&- \left(\frac{\partial F_{xi}}{\partial s_{Li}} - \frac{\partial F_{yi}}{\partial s_{Li}} \cdot \delta_i\right) b_i \\ &+ \left(\frac{\partial F_{xi}}{\partial s_{Li}} \cdot \delta_i + \frac{\partial F_{yi}}{\partial s_{Li}}\right) l_i \Bigg] \end{aligned} \tag{31}$$

with the derivatives of F_{xi}, F_{yi} to be determined from (5), (6). With Δs_{Ll} from (31) and ΔF_{x1}, ΔF_{x2} from (5) the required increases of the brake pressures result from (1):

$$\Delta p_i = \frac{\Delta M_{BLi}}{c_{pF,R}} = \frac{\Delta F_{xi} \cdot R}{c_{pF,R}} , \quad i = 1, 3 . \tag{32}$$

The determination of the steering increments $\Delta\delta_F$ or $\Delta\delta_R$ to provide ΔM_{yaw} is done in an analogous way to (31).

3.4 Actuator

As a model of the respective dynamic actuator behaviour a time delay represented by first or second order filters is applied.

The additional steering angle δ_i^A at the wheel i is determined by

$$T_\delta^2 \, \ddot{\delta}_i^A + 2D_\delta T_\delta \dot{\delta}_i^A + \delta_i^A = \bar{\delta}_i , \quad \bar{\delta}_i = \delta_{i0}^A + \Delta\delta_i , \quad i = F, R \tag{33}$$

with the values $T_\delta = 0.015$ and $D_\delta = 0.8$, the actual already applied additional steering angle δ_{i0}^A and the transformer output $\Delta\delta_i$. The additional steering angles δ_i^A are limited with 2°. Considering the steering angles $\delta_{i,d}$ due to driver's activities the steering angles at the wheels are finally

$$\delta_i = \delta_{i,d} + \delta_i^A , \quad i = 1..4 . \tag{34}$$

Similarly, the pressure p_i at wheel i is applied after the output Δp_i of the transformer passes a first-order filter

$$\dot{p}_i = \frac{1}{T_p}(\bar{p}_i - p_i) , \quad \bar{p}_i = p_{i0} + \Delta p_i \tag{35}$$

with the value $T_p = 0.01$ and the corresponding actual pressure p_{i0}. The pressure p_i which is to be applied by the brake system may be further modified according to the limitations set by the ABS-system.

4 RESULTS

To be able to compare the different control strategies and their possible interaction with the ABS-system, 'braking during cornering' is chosen as a relevant manoeuvre. The initial condition is steady-state cornering with a speed of $v_x = 25m/s$ and a lateral acceleration of $a_y \approx 4m/s^2$ ($\Rightarrow$ radius $\approx 156m$).
The high friction surface has a maximum force coefficient of $\mu_{max} \approx 0.9$. For the steady-state motion a fixed steering angle and accelerator input is determined. Then the accelerator is released at $0.1s$ and braking starts at $0.2s$. The brake

pedal activation is increased linearly form zero to its maximum within another 0.1s. The steering angle at the hand-wheel is kept constant.

Fig.4 shows the principle behaviour of a passenger car at the manoeuvre 'braking during cornering': If the deceleration is quite low the car turns inside and if the brake pedal activation is higher than a certain level it may leave the desired path to the outside.

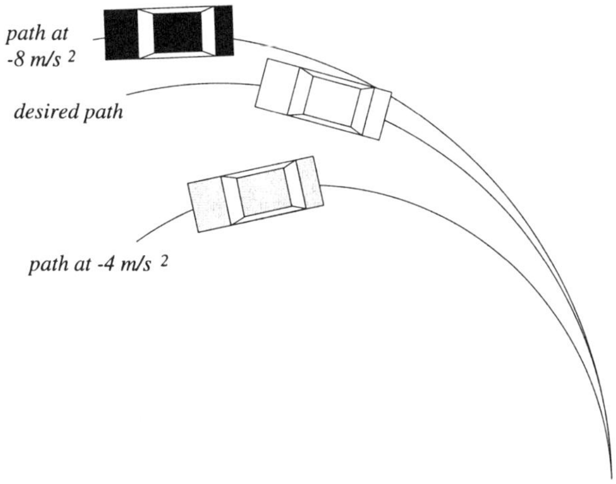

Fig. 4: Principal behaviour of a car at the manoeuvre 'braking during cornering'.

The desired or reference solutions in Figs.5,6,8 are the quasi steady-state lateral acceleration or yaw velocity derived with the 2-wheel vehicle model from equation (17) and $a_{q,d} = \dot{\psi}_d \cdot v_{ref}$.

Figs.5 and 6 show the behaviour of the car for low and high longitudinal deceleration, respectively. Without yaw moment control and a low deceleration ($\approx 4m/s^2$, Fig.5) the yaw velocity and the lateral acceleration increase which is corresponding to a deviation of the car from the desired path to the inside - see Fig.4. For a high deceleration ($\approx 8m/s^2$, Fig.6) yaw velocity and lateral acceleration of the car are smaller compared to the reference values. The car tends to leave the desired path tangentially - see Fig.4.

The comparison of the three different control strategies in Fig.5,6 demonstrates a good compensation at the low deceleration for all strategies but at the high deceleration manoeuvre only the two additional steering systems are able to influence the car in the desired way. The unilateral braking strategy tries to apply additional brake pressure to the less loaded inner wheels but the ABS limits the desired pressure increase to avoid wheel-locking.

In addition Fig.7 depicts the brake pressures p_i (normalized with respect to the maximum limit of the brake pressure) for all wheels corresponding to Figs.5,6 for the unilateral braking strategy. For low deceleration the pressure modulation of the outer wheels 2, 4 can be realized without activity of the ABS-system. But the high deceleration manoeuvre demands ABS-activities at the inner wheels. At the beginning of the braking the unilateral braking strategy increases the pressure at wheels 2, 4 for the short period of the deceleration build up. Although afterwards an increase of the braking forces at the inner wheels 1, 3 would be necessary, the ABS constrains their brake pressure. That is why no control effect can be achieved in this case.

For different longitudinal decelerations the relations $\dot{\psi}_{1s}/\dot{\psi}_0$ (yaw velocity $\dot{\psi}_{1s}$ one second after starting braking related to the initial steady-state yaw velocity $\dot{\psi}_0$) in Fig.8a and correspondingly $a_{y,1s}/a_{y,0}$ in Fig.8b for all considered control strategies and the uncontrolled case are compared.
Fig.8 once more shows that all control strategies work quite well until a deceleration of about $6.5m/s^2$. From then on the unilateral braked vehicle behaves similar to the uncontrolled one because of the brake-pressure limitations due to the ABS. With additional steering efficiency is lost too but the car is still influenced in the desired way at higher decelerations.

5 CONCLUSION

This paper covers three possibilities for actuation strategies for a yaw moment control of a passenger car for stabilizing a car in a critical driving situation - unilateral braking or additional steering of the front or the rear wheels. The characteristic manoeuvre 'braking during cornering' shows that the control system works quite well with all strategies for low decelerations on the one hand but only the additional steering strategies show an improvement of the vehicle behaviour at high decelerations on the other hand. In particular with respect to the ABS-system and the unilateral braking strategy there is a conflict between the desired deceleration by the driver and the controller action. There may be a better compensation of the path deviation by unilaterally reducing the brake pressure in some situations but that can also be problematic with respect to accident avoidance due to a longer stopping distance.

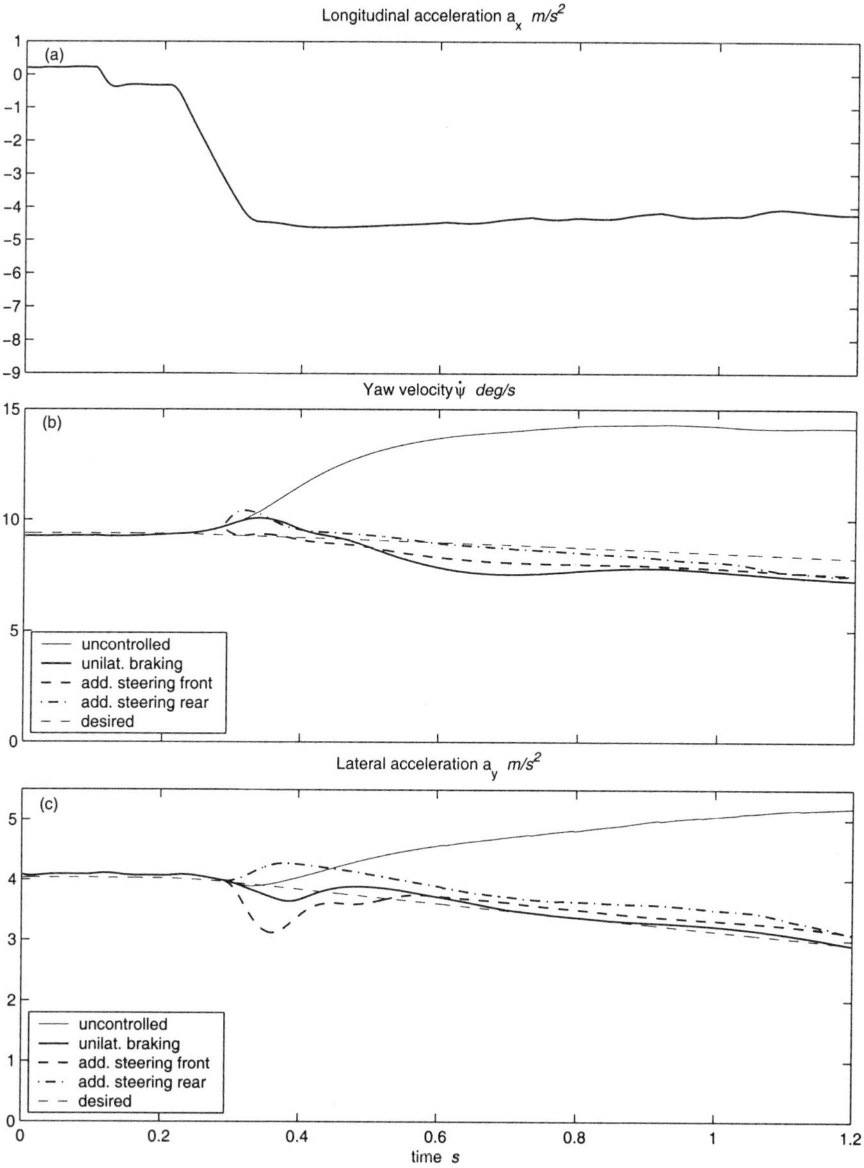

Fig. 5: Braking during cornering with $a_x \approx -4m/s^2$.

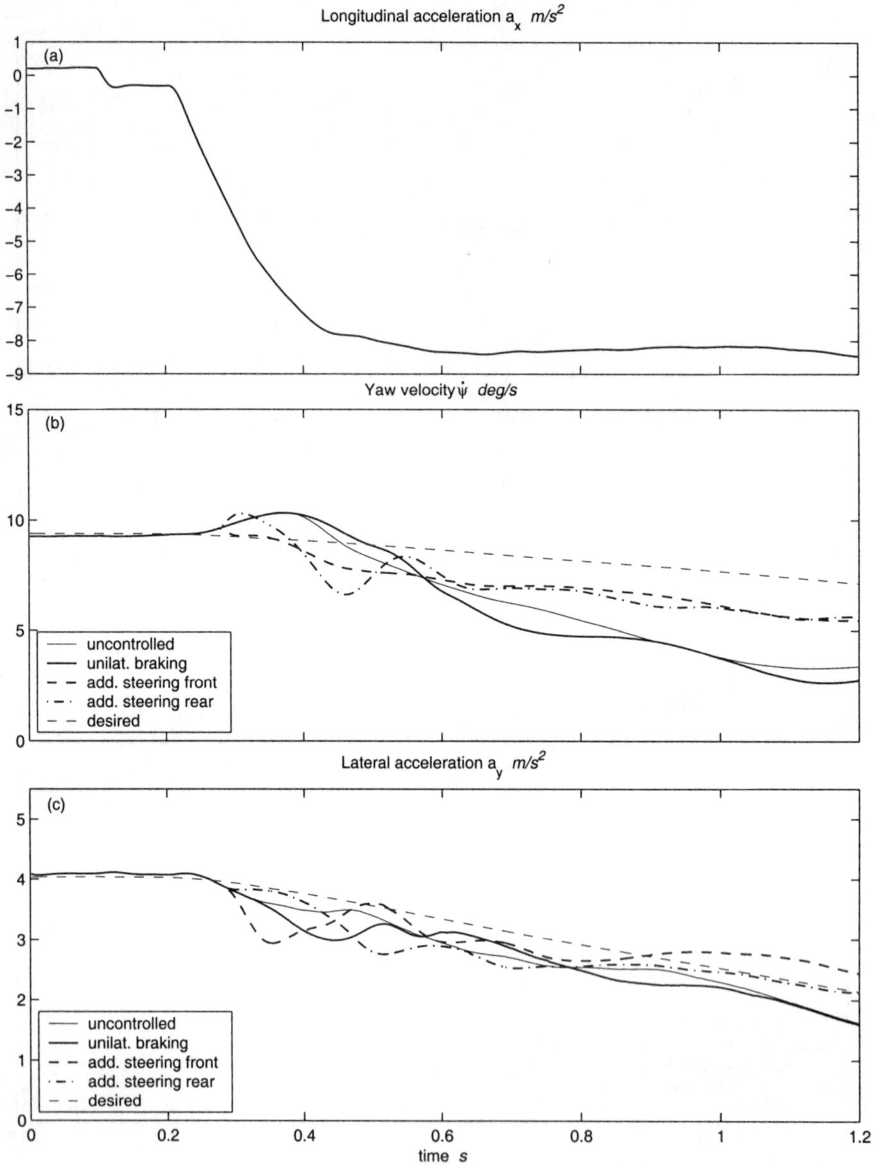

Fig. 6: Braking during cornering with $a_x \approx -8m/s^2$.

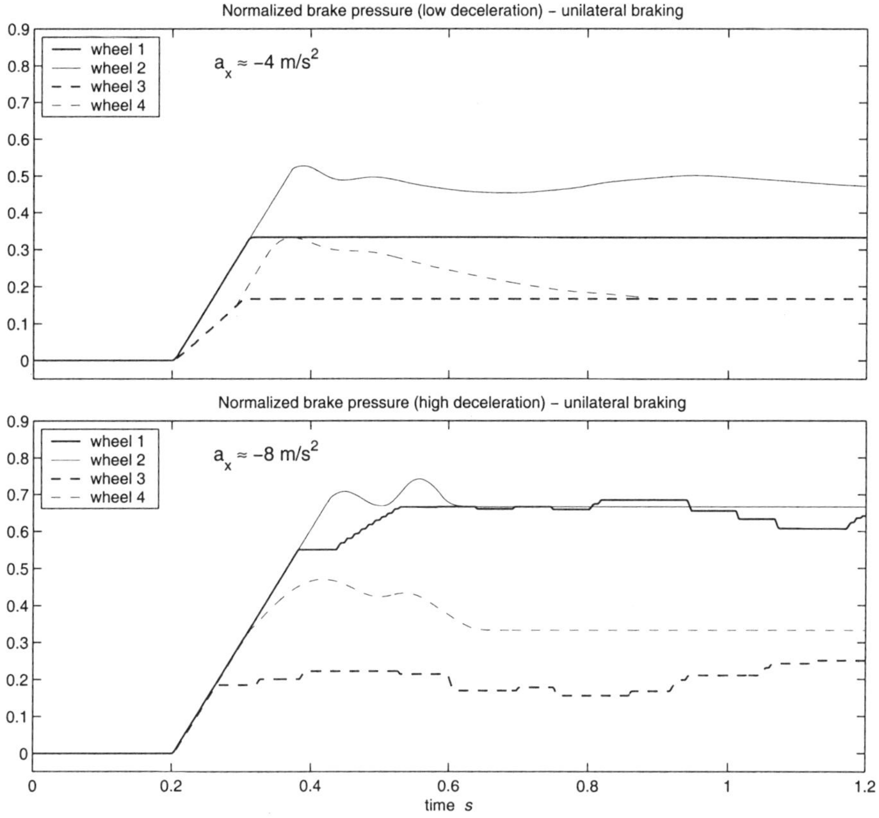

Fig. 7: Brake pressures corresponding to Fig.5,6 for unilateral braking.

REFERENCES

1. Lugner, P., Mittermayr, P.: Possibilities to Improve the Vehicle Cornering Dynamics by the Tyre Forces. Proc. of the 11th IAVSD Symposium, Swets & Zeitlinger, 1989.
2. Konik, D., Müller, R., Pestl, W., Toelge, T., Leffler, H.: Elektronisches Bremsenmanegement als erster Schritt zu einem integrierten Chassis Management. Teil 1, ATZ 101/4, 1999.
3. Nagai, M., Hirano, Y., Yamanoka, S.: Integrated Robust Control of Active Rear Wheel Steering and Direct Yaw Moment Control. Proc. of the 15th IAVSD Symposium, Swets & Zeitlinger, 1998.
4. Getting started using ADAMS/Car. Mechanical Dynamics, 1999.
5. Using ADAMS/Solver. Mechanical Dynamics, 1998.

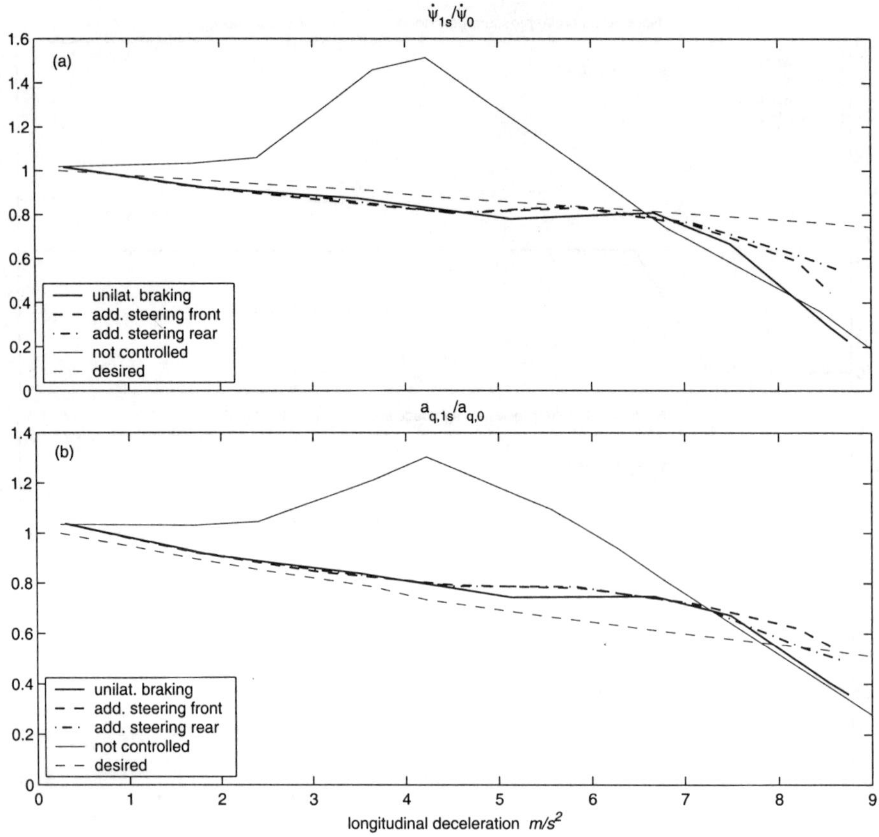

Fig. 8: Comparison of yaw velocity and lateral acceleration 1s after starting braking related to their steady-state values.

6. Lugner, P., Mittermayr, P.: A Measurement Based Tyre Characteristics Approximation. Proc. of the 1st International Colloquium on Tyre Models for Vehicle Dynamics Analysis, Swets & Zeitlinger, 1993.
7. Burckhardt, M.: Fahrwerktechnik: Radschlupf-Regelsysteme. Vogel Verlag, 1993.
8. v. Zanten, A., Erhardt, R., Pfaff, G.: FDR - Die Fahrdynamikregelung von Bosch. ATZ 96/11, 1994.
9. Ammon, D.: Modellbildung und Systementwicklung in der Fahrzeugdynamik. Teubner Verlag, 1997.
10. Kimbrough, S.: Coordinated Braking and Steering Control for Stop and Acceleration. DE-Vol. 40, Advanced Automative Technologies, ASME, 1991.
11. Lugner, P., Plöchl, M., Riepl, A.: Investigation of Passenger Car-Trailer Dynamics Controlled by Additional Braking of the Trailer. Proc. of the International Symposium on Advanced Vehicle Control, Vol 2., 1996.
12. Slotine, J.-J.E., Li, W.: Applied Nonlinear Control. Prentice Hall, 1991.
13. Fennel, H., Klusemann, R., Kranz, Th., Kümmel, M., Schmidt, R.: Bremsregelsysteme in Fahrzeugen mit Allradantrieb - Teil 2. ATZ 102/10, 2000.

Vehicle System Dynamics Supplement 35 (2001), pp. 101-117

Optimal Preview Car Steering Control

R. S. SHARP* and V. VALTETSIOTIS**

SUMMARY

Steering of a car to follow a prescribed path with minimal error is represented as a problem in optimal linear discrete time preview control. A standard yaw/sideslip linear car model is put into discrete time form and joined with a road preview model. A quadratic cost function consisting of terms describing path and attitude errors with respect to a roadway and steering angle control is minimised by standard optimal control theory methods, using MATLAB™ to perform the necessary computations. The form taken by the optimal preview control, with variations in the weighting of terms in the cost function and with variations in the vehicle speed, is established. The results are examined in the context of real driving and tested by simulation of manoeuvres using the optimal steering control in what is thought to approximate the manner of a real driver. New insights into steering of vehicles are generated and new avenues for exploration of the influences on the driver of varying the vehicle dynamic properties are opened up.

1. INTRODUCTION

Much effort has gone into the creation of driver models, mainly concerned with steering control. These are of value since they enable the simulation of path specified manoeuvres, as opposed to control input specified manoeuvres. They may contribute to improved understanding of the driving activity itself and of the way in which drivers arrive at judgements of vehicle quality. This, in turn, reflects on the very nature of vehicle quality, which engineers wish to build in to their designs. Further, automated steering control systems under investigation currently will be better realised if human driving is understood thoroughly first. Assisting the driver by electronically controlled chassis and powertrain systems is a worthy aim but it had better be done with a realistic appreciation of the way in which the driver operates.

Driver models were reviewed in [1]. Attention was concentrated on models depending on single point preview of the path ahead, with more complex processing of that basic information notionally providing the means for performance improvement. An update and a perspective on this work was given in [2], in which an association was made between linear optimal preview control and road vehicle steering. In that paper, the known structure of the linear optimal preview controller and the known manner in which its parameter values relate to the dynamics of the controlled system [3, 4] were used to create a driver steering model. The basic structure of the controller is illustrated in Fig. 1. The driver is represented as acquiring many sample values of the path preview error in front of the vehicle and forming a weighted sum of these values as a contribution to the steering angle input.

The weights are chosen to reflect (a) the exponential decay in the use of preview information with preview distance, inherent in the linear optimal control solution and (b) the dynamic properties of the vehicle. To deal with tyre force saturation, which sets the limit to a vehicle's lateral performance, control gain saturation features were added in a heuristic fashion and realistic upper and lower limits were put on the steer angle. Excellent path tracking performance was demonstrated, together with unusually robust properties of the controller, such that driving at racing speeds could be simulated and very little tuning of the controller parameters to changes in the vehicle design were necessary. Cars with completely opposite front / rear force balances could be "driven" in a realistic fashion by the same controller.

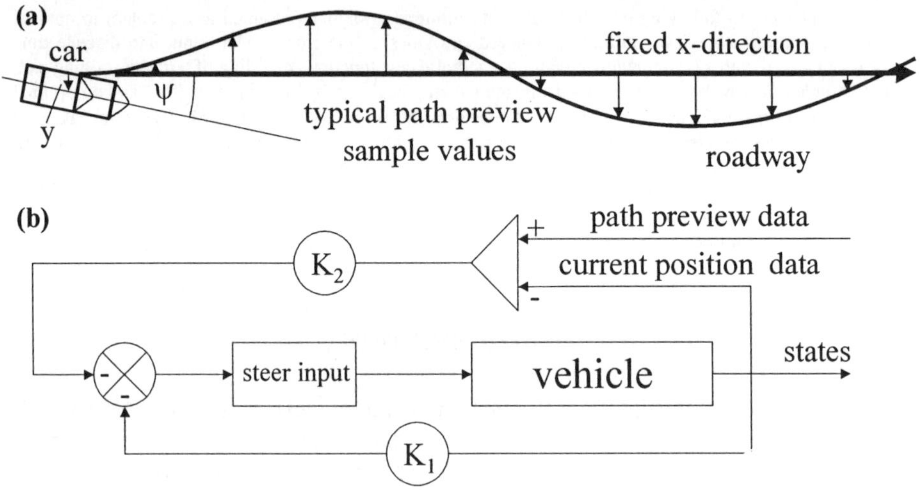

Fig. 1. Preview data acquisition (a) and linear optimal discrete-time controller structure (b).

A clear need for a formal solution of the linear optimal steering control problem arises from this prior work. The problem formulation devised by Louam [5] and improved by Prokop [3] provides a methodology for setting up the driver / vehicle preview control problem, such that it conforms to the requirements of standard discrete time linear quadratic regulator theory [6]. This formulation, including a simple vehicle and a randomly curved road representation, is utilised and tested against the requirements of the theory. The optimal control problem is then solved, the solution method being described. Solutions are interpreted and tested through simulated path following, to ensure that the control is effective. The basic nature of the optimal steering controller is revealed. Variations in system behaviour with changes in the relative costs of path errors, attitude errors and steering control inputs are examined. Influences of vehicle speed are also examined and interpreted.

2. THE VEHICLE MODEL

The vehicle model used is the standard yaw / sideslip one, an account of which is given in [7]. The car is represented as a single rigid body, moving on a level plane at constant forward speed. It has freedom to rotate in yaw and to translate laterally. It contains a rigid linkage steering system, such that turning the steering wheel through a certain angle steers the front road wheels through a proportionate angle, but the inertial effects of steering the wheels are not included. Aerodynamic forces are not included on the basis that they are relatively small at normal speeds for normal cars. Tyre lateral forces are modelled as proportional to slip angle, while aligning moments are discounted.

Since vehicles are substantially symmetrical, traction forces are usually evenly divided through differential gear action and the forward velocity of any point is normally dominated by the forward velocity of the whole car, it is as if the two wheels of an axle were combined in the centre-plane. Small differences in running conditions between left and right sides of the vehicle are ignored. The model is sometimes described as a "bicycle" model, although it in no way represents a bicycle, for which roll and steer freedoms are vital to a good representation [8]. The car model is illustrated in Fig. 2, where tyre forces, dimensions and coordinates are shown.

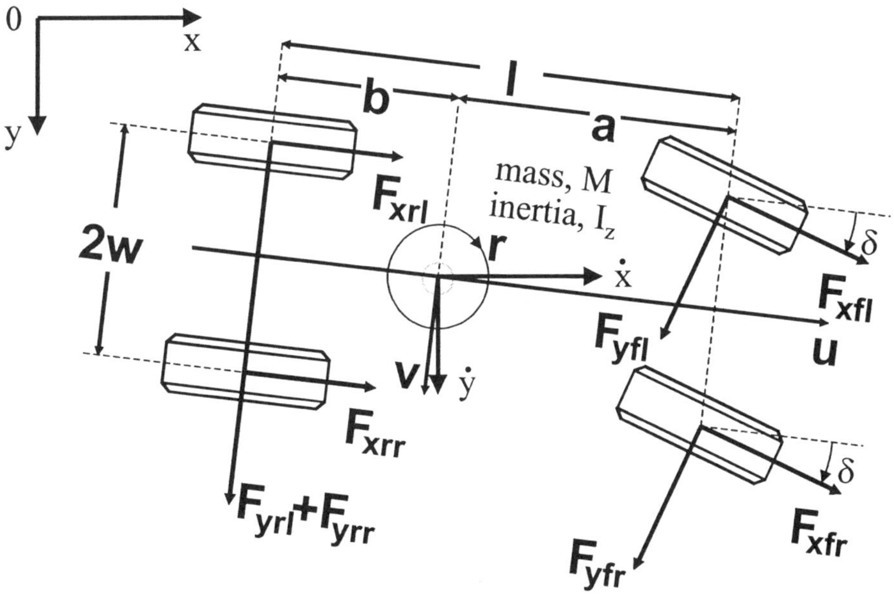

Fig. 2. Plan view of car, showing forces, key dimensions and coordinates.

Tyre sideslip angles can easily be deduced from Fig. 2 to be:

$$\alpha_f = (\dot{y} + a\dot{\psi})/u - \psi - \delta_{sw}/G \qquad and \qquad \alpha_r = (\dot{y} - b\dot{\psi})/u - \psi$$

in which G is the steering gear ratio, relating hand wheel angle to road wheel angle, from which the equations of motion follow straightforwardly. Employing the notations:

$$X_1 = y, \quad X_2 = \dot{y}, \quad X_3 = \psi, \quad X_4 = \dot{\psi}$$

and writing the equations in standard state-space form, $\dot{\mathbf{X}} = \mathbf{A}.\mathbf{X} + \mathbf{B}.\delta_{sw}$, then:

$$\mathbf{A} = \begin{bmatrix} 0 & 1 & 0 & 0 \\ 0 & -(C_f + C_r)/Mu & (C_f + C_r)/M & (bC_r - aC_f)/Mu \\ 0 & 0 & 0 & 1 \\ 0 & (bC_r - aC_f)/I_z u & (aC_f - bC_r)/I_z & -(a^2C_f + b^2C_r)/I_z u \end{bmatrix} \quad \textit{and} \quad \mathbf{B} = \begin{bmatrix} 0 \\ C_f/MG \\ 0 \\ aC_f/I_z G \end{bmatrix}$$

Transformation of the equations of motion into discrete time form is a standard process [6] that can be done automatically in MATLAB [9]. The equation form is then: $\mathbf{X}(k+1) = \mathbf{A_d}.X(k) + \mathbf{B_d}.\delta_{sw}(k)$, in which k is the index of time; that is, the elapsed time from the chosen zero to the present time is kT, T being the sampling interval. The steering wheel input, δ_{sw}, is a zero order hold process in this representation.

3. THE ROAD PREVIEW MODEL

The roadway is described by its lateral profile. That is, the lateral deviation, $\mathbf{y_r}$, from a fixed straight line, the x-axis, is given as a discrete function of x, such that it is defined at the sampling times kT, corresponding to a specified vehicle forward speed.

At an instant k, the road preview data observed by the driver consists of say (h+1) y_r values at and ahead of the car equally spaced in x, Fig. 3. $\mathbf{y_r} = [y_{r0}\ y_{r1}\ y_{r2}\ y_{rh}]^T$. The x-spacing is uT, where u is the constant car speed. At the next instant, (k+1), the first road preview sample (at the x-location of the car reference point) from time k is lost from the problem, the second from k becomes the first at (k+1) and so on. The sample value h at time k becomes the value (h-1) at instant (k+1) and a new value, previously outside the problem, enters as value h at (k+1). It is convenient to regard this as the single external input to the system and to regard the other h sample values as system states [3]. The road preview updating process involves a shift register type operation in going from k to (k+1) in time. Mathematically it is described as in [5, 3] by: $\mathbf{y_r}(k+1)=\mathbf{D}.\mathbf{y_r}(k) + \mathbf{E}.y_{ri}$, in which $\mathbf{y_r}$ is a vector of length (h+1), y_{ri} is the input to the road system, and

$$\mathbf{D} = \begin{bmatrix} 0 & 1 & 0 & . & . & . & 0 \\ 0 & 0 & 1 & . & . & . & 0 \\ . & & & & & & . \\ . & & & & & & . \\ . & & & & & & . \\ 0 & 0 & 0 & . & . & . & 1 \\ 0 & 0 & 0 & . & . & . & 0 \end{bmatrix} \quad \textit{and} \quad \mathbf{E} = \begin{bmatrix} 0 \\ 0 \\ . \\ . \\ . \\ 0 \\ 1 \end{bmatrix}$$

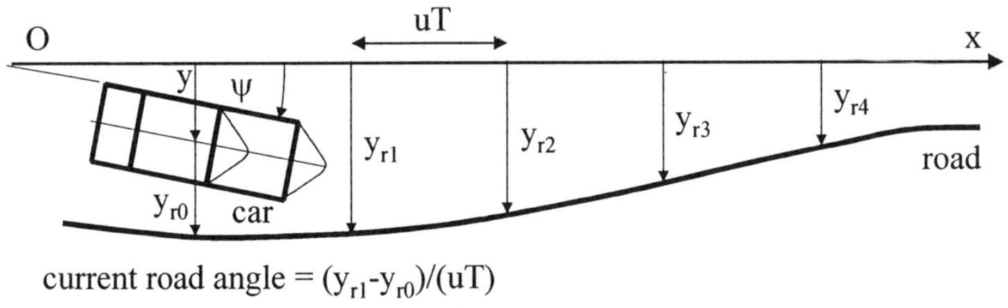

Fig. 3. Snapshot of car and road at time instant k.

4. THE OPTIMAL CONTROL PROBLEM AND ITS FORMAL SOLUTION

Combining the discrete time equations for the car with those for the road, we have:

$$\begin{bmatrix} \mathbf{X}(k+1) \\ \mathbf{y_r}(k+1) \end{bmatrix} = \begin{bmatrix} \mathbf{A_d} & \mathbf{0} \\ \mathbf{0} & \mathbf{D} \end{bmatrix} \begin{bmatrix} \mathbf{X}(k) \\ \mathbf{y_r}(k) \end{bmatrix} + \begin{bmatrix} \mathbf{0} \\ \mathbf{E} \end{bmatrix} y_{ri} + \begin{bmatrix} \mathbf{B_d} \\ \mathbf{0} \end{bmatrix} \delta_{sw}$$

with no connection, so far, between car and road. The connection will come from specifying the cost function for the optimisation.

Following Louam [5] and Prokop [3], we seek the time invariant optimal control which minimises the expected value of the quadratic cost function, under white noise excitation. The cost is written in the form [3]:

$$J = \lim_{n \longrightarrow \infty} \sum_{k=0}^{n} \{(\mathbf{Z}^T(k).\mathbf{R_1}.\mathbf{Z}(k) + \delta_{sw}(k).R_2.\delta_{sw}(k))\} \ \ with \ \ \mathbf{Z} = [\mathbf{X} \ \ \mathbf{y_r}]^T$$

with $\mathbf{R_1}$ chosen to reflect the system objectives as well as possible. $\mathbf{R_1}$ must be positive semi-definite and R_2 must be a positive scalar. The main purpose is path following, the quality of which can be specified by the sum of the squares of the differences between the y-coordinate of the car's reference point and the corresponding value y_r for the lateral displacement of the road. Also, we may wish to minimise the corresponding sum of squares of attitude angle differences, a typical term being $[\psi(k)-\{y_{r1}(k)-y_{r0}(k)\}/(uT)]^2$. The first term here represents the attitude angle of the car at time instant k, while the second term is a simple version of the road angle at the current x-coordinate of the car. These priorities are reflected by setting:

$$\mathbf{R_1} = \mathbf{C^T Q C} \ \ with \ \ \mathbf{C} = \begin{bmatrix} 1 & 0 & 0 & 0 & -1 & 0 & 0 & 0 & \ldots\ldots & 0 \\ 0 & 0 & 1 & 0 & 1/(uT) & -1/(uT) & 0 & 0 & \ldots\ldots & 0 \end{bmatrix}$$

$$and \ \ \mathbf{Q} = \begin{bmatrix} q_1 & 0 \\ 0 & q_2 \end{bmatrix} \ \ corresponding \ to \ the \ state \ vector:$$

$\mathbf{Z} = [y \ \ \dot{y} \ \ \psi \ \ \dot{\psi} \ \ y_{r0} \ \ y_{r1} \ \ y_{r2} \ \ y_{r3} \ \ \ldots\ldots \ \ y_{rh}]^T$, with R_2=1.

The relative importance attached to path errors, attitude errors and steer angle are set by choosing q_1 and q_2 appropriately. The system structure is illustrated in Fig. 4, which shows how the car and road sub-systems come together through the cost function. The structure is a subset of that of [3] involving preview suspension control. The difference arises from the fact that, in the suspension problem, the road elevation is first observed by the preview sensor and then it provides the disturbance input to the suspension, while, in the driving problem, the lateral road contour is observed by the driver only. It does not provide any disturbance directly to the car.

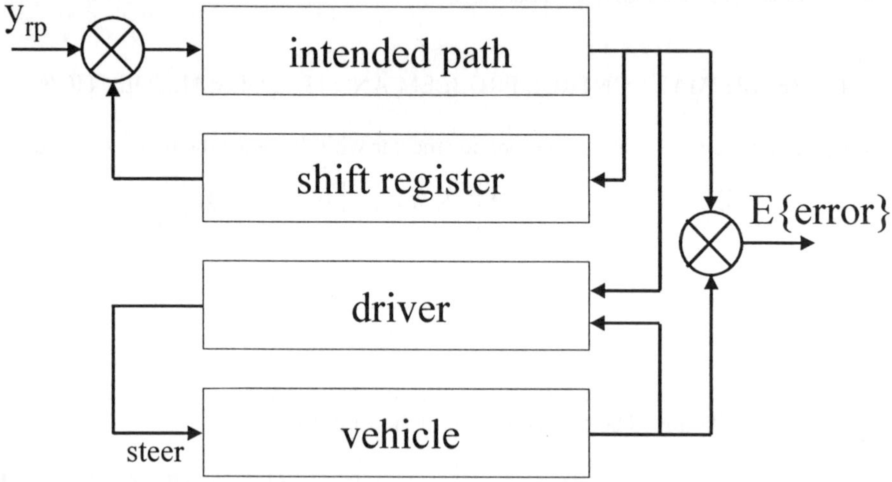

Fig. 4. Car / road system structure with optimal preview control formulation.

The time-invariant optimal control, minimising J, is of the form: $\delta_{sw}(k) = -\mathbf{K}.\mathbf{Z}(k)$. It was shown first by Louam [5] and confirmed by Prokop [3] that the control parameters can be determined by initially solving the non-preview problem for the part of **K**, (**K**$_1$ say) representing state feedback. Then that solution can be used to determine the remainder of **K**, (**K**$_2$ say), representing the preview control. Then: $\delta_{sw}(k) = -\mathbf{K}.\mathbf{Z}(k) = -[\mathbf{K}_1 \quad \mathbf{K}_2][\mathbf{X} \quad \mathbf{y}_r]^T$. The detailed theory is given in precisely the form required in [3] and it is not repeated here. The necessary conditions for the control calculations to work, concerning stabilisability and detectability, are satisfied here exactly as in [3]. The detailed theory of [3] is, of course, utilised in the MATLAB coding, by which the numerical calculations are performed.

5. PARAMETER VALUES, CALCULATION DETAILS AND CONTROL LAW RESULTS

The simple vehicle model requires only very basic parameters to define the car. The baseline case is defined in Table 1. This baseline vehicle has $aC_f = bC_r$, so that it is neutral steering [7].

The optimal control law parameters are found by the following sequence, in MATLAB. First, a program "car_par", containing the car parameters above, is run. Then "car_eq" forms the equations of motion in state space form, given the forward speed, and uses the MATLAB continuous to discrete transformation function, C2D, to convert the problem to discrete time form with a time step of 0.02s. A program "build" then uses a routine "cost", containing the values of q_1 and q_2, the weights, and calculates the time-invariant optimal non-preview control using the MATLAB function DLQR. This control, $\mathbf{K_1}$, is then used in the computation of $\mathbf{K_2}$, the preview gain vector, via the number of preview samples available to the driver. By changing the priorities in the cost function, many different controls can be derived. Here, attention will be paid to just three cases. In the first of these, the priority is on path following, with $q_1=100$ and $q_2=0$. In the second, the objective is keeping the car tangential to the track, with $q_1=0$, $q_2=100$. In the third, accurate path following with only modest use of the steering control input is the aim, as reflected by $q_1=0.05$, $q_2=0$. It should be remembered that, in each case, R_2, associated with the steer angle input, is set to unity.

Table 1. Baseline vehicle parameter set.

Mass of whole car, kg	$M = 1200$
Yaw inertia, kgm^2	$I_z = 1500$
Mass centre to front axle distance, m	$a = 0.92$
Mass centre to rear axle distance, m	$b = 1.38$
Cornering stiffness of front axle tyres, N/rad	$C_f = 120000$
Cornering stiffness of rear axle tyres, N/rad	$C_r = 80000$
Steering gear ratio; hand wheel/road wheels	$G = 17$

The feedback gains $K_1(2)$ and $K_1(4)$, relating to lateral velocity and yaw rate respectively are shown in Fig. 5, while the corresponding preview control gains are shown in Figs 6, 7 and 8. The feedback gains $K_1(1)$ and $K_1(3)$ are not shown, since these are part of the global problem but not part of the local one, see section 6. They are derivable from the preview gains, as argued below.

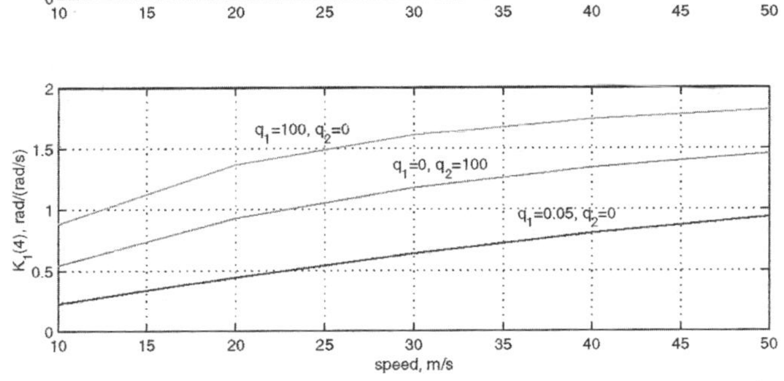

Fig. 5. Optimal feedback gains relating lateral velocity and yaw rate to steering wheel angle as functions of speed, for different cost weightings.

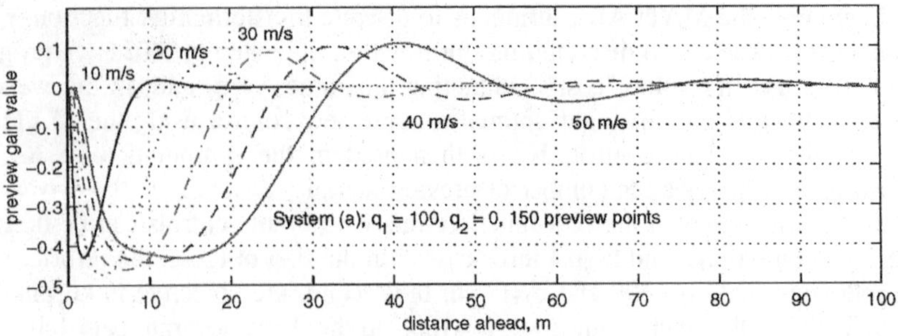

Fig. 6. Optimal preview gains for five speeds; 150 preview points, with q_1=100 and q_2=0; system (a).

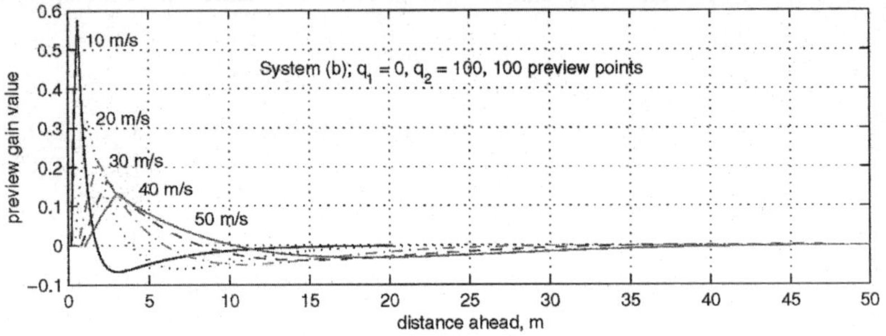

Fig. 7. Optimal preview gains for five speeds; 100 preview points, with q_1=0 and q_2=100; system (b).

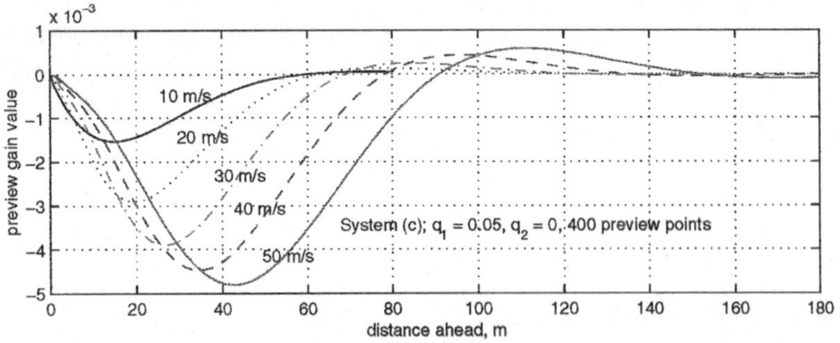

Fig. 8. Optimal preview gains for five speeds; 400 preview points, with q_1=0.05 and q_2=0; system (c).

6. INTERPRETATION OF THE OPTIMAL CONTROL LAWS

Understanding the relationship between these optimal gains and real driving requires considerable thought, since the theory relied upon has used a fixed reference system view of events in order to fit with prior results, while, in reality, all the path observations are made by the driver from his/her moving position. In the theory, there is a fixed x-axis, a moving vehicle with a known positional relationship to the x-axis and driver knowledge of the path ahead in an absolute sense. Updating of the path preview data involves a shift register type of operation. In reality, the driver has no basis or need for thinking about a fixed x-direction but rather will keep track, in a continuous manner, of the relationship between the car and the road ahead. That is, the driver is concerned with relatives and not absolutes.

It appears that the driver will effectively position the x-axis anew at each time step, to align with the current position of the vehicle, Fig. 9. The road preview information is then sampled in a way similar to but independent of the previous samplings. There is no absolute lateral position or attitude angle error, since the x-axis is re-located by the driver to make it so. The optimal steering input, for a given path following task, must be the same in the absolute world of the theory and in the relative world of the driver. In shifting from the global view to the local view, the consequences for the information contained in the control law can be seen. Certain relationships between the terms in the optimal control are implied, as will now be demonstrated.

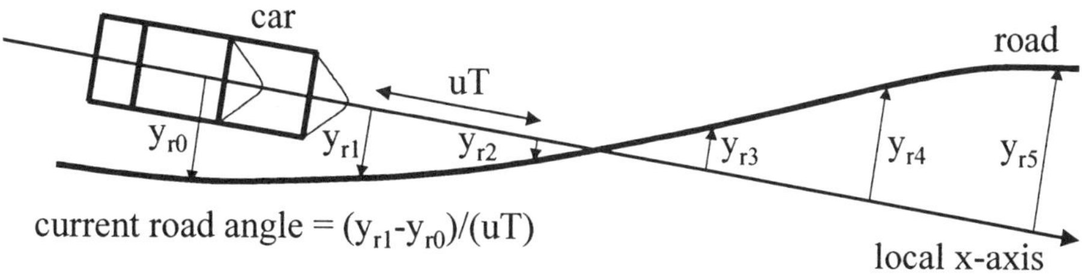

Fig. 9. Optimal preview steering control in the local view of the driver. Note the absence of lateral position and attitude angle errors in the local view.

First, let us consider the translation of the x-axis in the y-direction from its original position such that it passes through the car reference point. The global y value is reduced to zero in the local system. The term $K_1(1)*y$ is lost from the feedback control. Also, the path preview error values each have the previous global y value subtracted from them. The steer angle can only remain invariant under the change if:

$$K_1(1) = -\sum_{i=0}^{h} K_2(i)$$

which turns out to be true of the numerical results generated. In a similar way, we expect invariance of the optimal steering angle control under the rotational shift implied by changing the global picture of the mathematical theory to the local picture of the driver. As a consequence of this change, the term $K_1(3)*\psi$ is lost from the feedback control, the lateral velocity changes its definition, since it is measured in a different direction after the change, and the ith preview sample value is reduced by $(i-1)uT\psi$, assuming small angles (it should be noted that y_{r0}, the first preview sample, corresponds with i=1). The three sets of changes must compensate each other, which requires that:

$$u*K_1(2)+K_1(3) = -\sum_{i=0}^{h}(i-1)*u*T*K_2(i)$$

which also turns out to be the case in the numerical results. As pointed out previously, it appears that the lateral position feedback and attitude angle feedback gains in the global optimal control are hypothetical. They come out of the process of treating the control problem as different, for analytical convenience, from how it really is. With the driver effectively re-positioning the reference line, the x-axis of the theory, at each time step, the absolute lateral position and attitude angle errors of the feedback controller are automatically zero at each calculation step and it follows that the gains associated with these variables have no influence. On the other hand, the lateral and yaw velocity feedback terms appear in the real problem and are a genuine part of the optimal steering control law.

The steering wheel angle to yaw rate feedback gain does not vary so much with the performance priorities or the speed, Fig. 5, but the lateral velocity gain is zero if q_1 is zero and it increases systematically as q_1 increases. In principle, a steer by wire car could have these feedback controls built into its design, relieving the driver of the need to do this part of the control. In practice, the yaw rate part of the control would be realisable without too much difficulty but the lateral velocity is difficult to measure and the gain is highly variable anyway, depending on the driver's preferences.

There is no effective difference between the road preview data of Figs 6, 7 and 8 and that of Fig. 3, so it can be concluded that the preview gains directly indicate how the time invariant optimal driver uses the path preview information. For the simple but representative car considered, the controls vary markedly with the driver's priorities. If the onus is on precise path following with little cost associated with use of the steering input, 1s preview time is all that is needed at low speeds and 1.5s is sufficient at high speeds. The "driver" pays most attention to the sample values just ahead of the car, using the perceived path error here to steer the "correct" way. That is, if the path ahead is to the right of the driver's optical lever, he/she uses this part of the information to steer to the right. Particularly at higher speeds though, the preview gain sequence is oscillatory. Certain sections of the previewed path are used to promote steering the "wrong" way.

The preview gain sequence can be seen, to some extent, as the inverse of the vehicle dynamics, the driver utilising this inverse to yield the steering control which will give the desired response. The more oscillatory are the vehicle dynamics, the more complex the preview gain sequence is likely to be. It will be seen later how the oscillatory pattern can give rise to an oscillatory steering input when a single change of direction is needed.

If the priority is placed on attitude angle control, with little cost on steering input activity, the optimal preview gain sequence changes dramatically, Fig. 7. Shorter previews are sufficient in these cases and there is less "oscillation" in the sequences. When a relatively high cost is put on use of the steering input and path following is the aim, much more preview is needed. About 8s look ahead is necessary at 50m/s for the conditions of Fig. 8. Again, it is true that somewhat less preview time is required for lower speeds. The preview gain pattern is not dissimilar to that for Fig. 6 but the values are much less. The differences between these gain sequences for different cost function definitions highlight the potential conflict from the driver's viewpoint between steering for best path following and steering for best attitude control.

The more oscillatory pattern of the gains for higher speeds is not unexpected, since the plant becomes more oscillatory for higher speeds. Exactly the same feature has been observed in the preview suspension problem [3, 4].

7. SIMULATED PATH FOLLOWING WITH THE OPTIMAL CONTROLLER

To check the proper operation of the optimally controlled system and to see how it works, road sections have been generated and the preview controlled vehicle has been simulated, travelling along these roads, in the following manner. Firstly, for a generally curved but realistic path, a random number sequence was created with RAND (in MATLAB). This was then scaled and an offset added to give a zero mean value to the sequence. The sample values were associated with successive positions along the road at 0.2m intervals (wavenumber = 5 cycles/m) and the sequence was smoothed with a 5th order Butterworth low pass filter with wavenumber 0.004 times that for the original data. A certain amount of trial and error was needed to establish this process, the resulting "road" appearing reasonable for its purpose. Secondly, a lane change manoeuvre was defined. This consisted of a cosine shaped 3.5m lateral shift in a 60m roadway length, joining two straight and parallel sections.

The absolute road model and the controlled car model were assembled and the MATLAB function DLSIM was used to simulate the car following each of these two roadways, with its steering input being derived mainly from the road preview. The path following was shown to be excellent [10].

Then the simulation program was re-written to represent the events taking place in the local view of the driver, Fig. 9, and a sudden 15 degree direction change manoeuvre was added to provide a third type of test.

As in the global case, the road was defined absolutely first. The car was then set up precisely on the path and precisely aligned with it, having no lateral position and no attitude angle errors. The computations involved outer and inner loops. In the outer loop, the car lateral velocity and yaw rate initial conditions were set (initially made zero) and the closed loop system matrix and the input distribution vector, formed as in the absolute case, were used to transform the state through one discrete time step.

The steering input and the lateral acceleration response were computed so that they could be output and examined. In the inner loop, the motion of the car was used to transform the road state information within the preview length, to represent the driver's view of it - the local view. The car state vector was augmented with all except the newest of these values. That newest value was the input scalar for the next step in time, allowing the next step to be taken. All the subsequent steps followed in the same way.

Each of the three road profiles described above has been tracked. The first of these is the random road, for which the results are identical with those coming from the DLSIM (global view) simulation. The second is the lane change manoeuvre, while the third involves the sudden 15 degrees change of direction.

Tracking results show the road path and the vehicle mass centre path overlaid, the path slope and the vehicle attitude angle overlaid, the steering wheel angle and the lateral acceleration of the car for a number of cases. In the various cases, different priorities, as above, were given to path following, attitude angle error minimisation and steer input minimisation.

Firstly, system (a) tracks a random road at 30m/s, Fig. 10, does the lane change at 30m/s, Fig. 11, and the direction change at 10m/s, Fig.12.

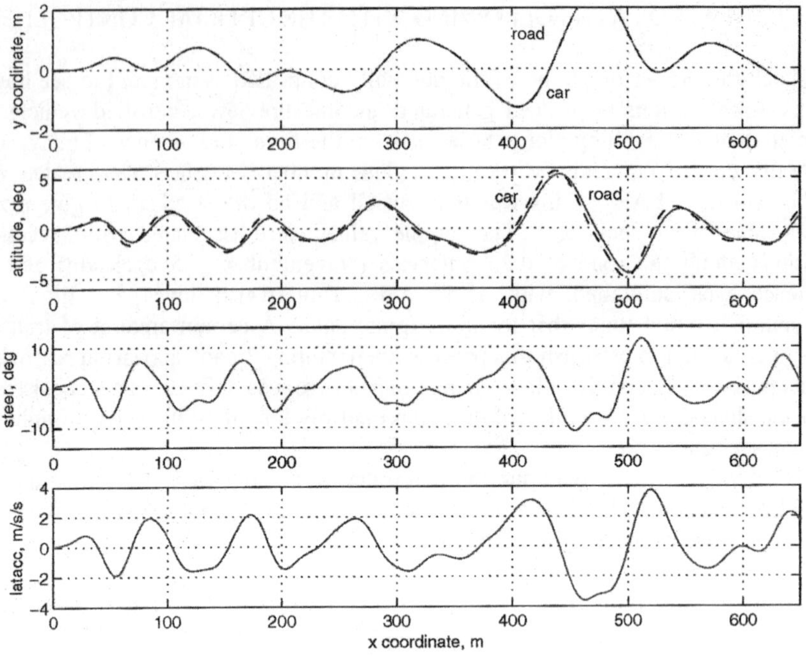

Fig. 10. Following a randomly generated path at 30 m/s with system (a).

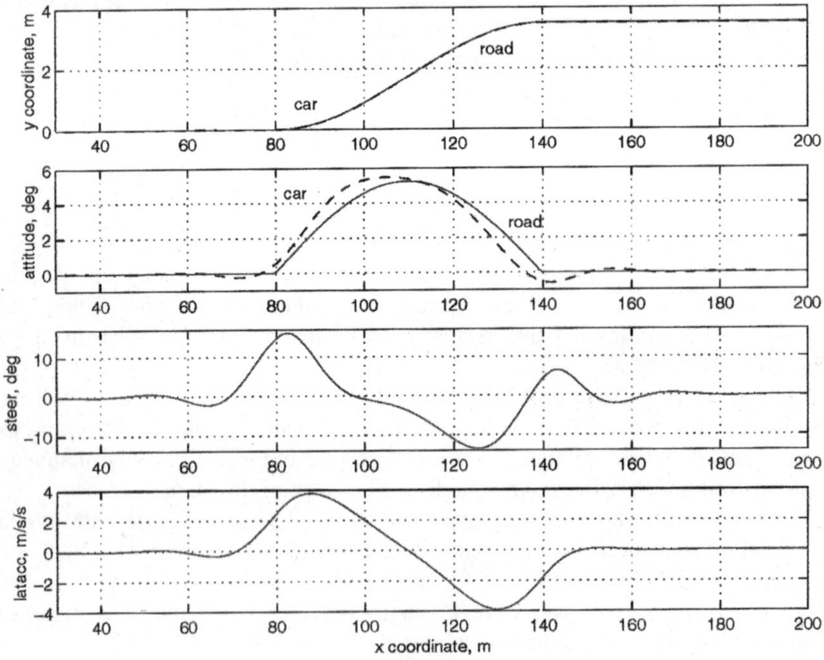

Fig. 11. Lane change manoeuvre at 30 m/s with system (a).

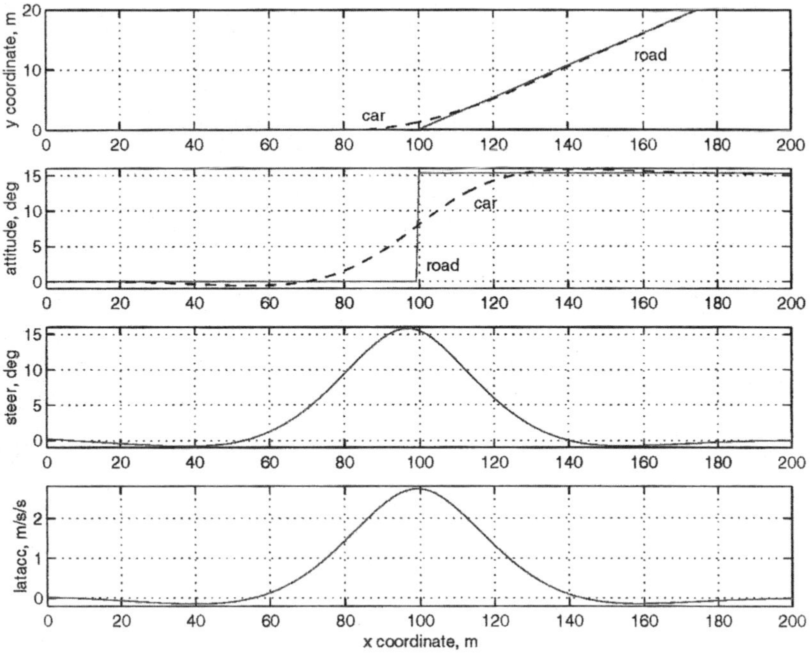

Fig. 12. Direction change at 10 m/s with system (a).

Then system (b), with its priority on attitude angle control, carried out the lane change at 30m/s, Fig. 13, and the direction change at 10m/s, Fig. 14. Finally, system (c) with its priorities divided between path tracking and economy of steering input, performed the direction change manoeuvre at 20m/s, Fig. 15.

In each of the simulations, the car speed was chosen to give a peak lateral acceleration which is compatible with the linearity assumption built into the whole of the theory.

The results demonstrate the excellent path following and attitude control, with the performance matching well the prioritisation in the control design. Oscillatory steering ahead of the lane change can be observed in Fig. 11, which also shows that the near-perfect path tracking implies some discrepancy between the path tangent and the car orientation.

Fig. 12, for the direction change, shows similar features, even though the speed is quite low. When there is a conflict between the path following and the attitude control, it can be surmised that a real driver will place the priority on the path following, so that system (a) appears much more important than system (b). System (c) was able to negotiate the direction change at 20m/s without needing very high lateral accelerations, due to the smoothing action of its long preview.

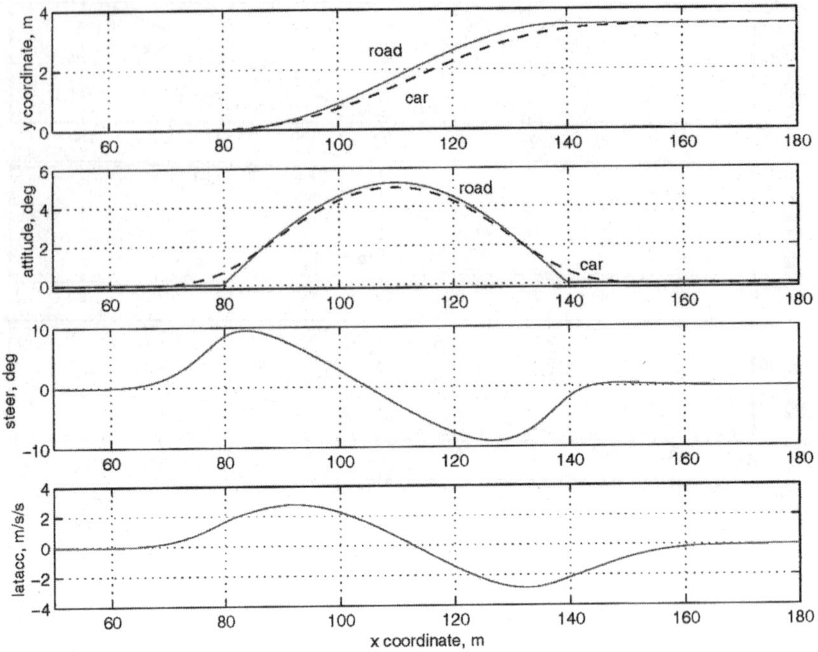

Fig. 13. Lane change manoeuvre at 30 m/s with system (b).

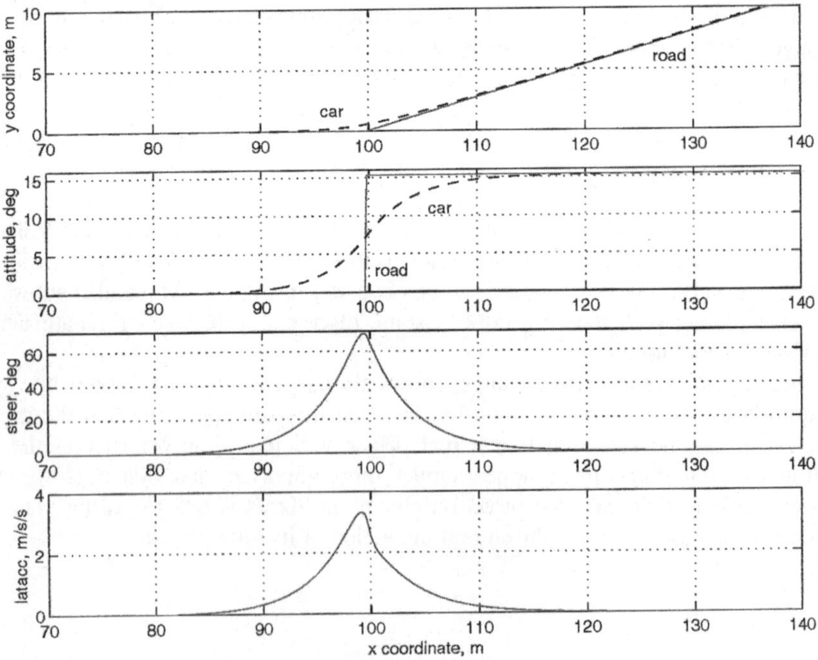

Fig. 14. Direction change at 10 m/s with system (b).

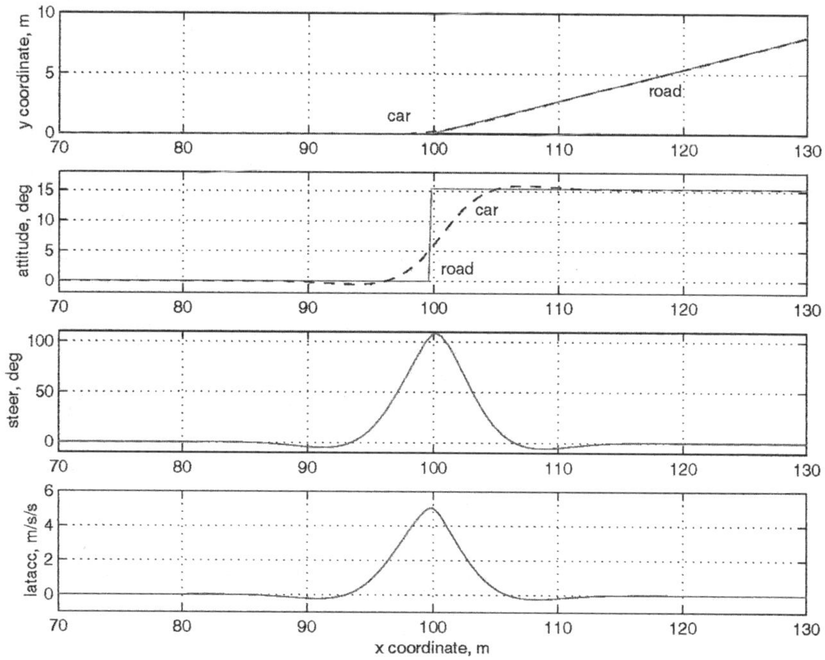

Fig. 15. Direction change manoeuvre at 20 m/s with system (c).

8. CONCLUSIONS

A new structure for the representation of optimal linear car steering control has been devised. The structure uses previous work in suspension control and it takes essentially the same mathematical form. The nature of the time invariant optimal preview and feedback control laws has been revealed and cases for a simple neutral steering car at different speeds and with different objectives have been solved. The theoretical results, generated first in an absolute framework, have been transformed to the relative framework of the driver. Since a real driver optimises by learning, it is expected that the new structure reveals, to some extent, how real driving works in the linear range. The preview control appears particularly significant and interesting.

The preview control shows the diminishing returns feature of previous studies and the car driver needs 1 to 1.5s preview time, depending on the travel speed, for path error minimisation. For attitude angle minimisation, he / she needs substantially less preview but if a high cost is placed on use of the steering input, much longer previews are needed for full benefit. The preview controls for path optimisation and heading optimisation are very different from each other and each changes with speed significantly. The preview gain sequence can be associated with the inverse dynamics of the vehicle, learned automatically by a human driver to give him / her a predictive capability. As the car travels faster, its yaw / sideslip mode damping factor decreases. It becomes more oscillatory [7]. Correspondingly, the inverse dynamics become more complex and the optimal preview control gain sequence becomes more "oscillatory" also. The gain pattern suggests that, at high speed, the driver may well use oscillatory steering control ahead of a turn to one side, as observed commonly in the case of rally cars negotiating small radius turns in one direction only.

Simulated path following has confirmed the capability of the controlled car to precisely track paths containing reasonable curvatures, if designed for this purpose. If very sharp changes of direction are demanded, interesting departures from perfect tracking occur. Systems designed for path control show attitude deviations and conversely. The driver apparently will suffer some conflict in deciding how to use the steering control to achieve his / her objectives, since both path and attitude errors cannot be made zero, with a conventional vehicle, by the same control.

REFERENCES

1. Guo, K. and Guan, H.: Modelling of driver/vehicle directional control system, Vehicle System Dynamics, **22**(3-4), 1993, 141-184.
2. Sharp, R. S., Casanova D. and Symonds P.: A mathematical model for driver steering control, with design, tuning and performance results, Vehicle System Dynamics, **33**(5), 2000, 289-326.
3. Prokop, G. and Sharp, R. S.: Performance enhancement of limited bandwidth active automotive suspensions by road preview, IEE Proceedings - Control Theory and Applications, **142**(2), 1995, 140-148.
4. Sharp, R. S.: Preview control of active suspensions, Smart Vehicles (J. P. Pauwelussen, H. B. Pacejka eds), Swets and Zeitlinger, Lisse, 1995, 166-182.
5. Louam, N., Wilson, D. A. and Sharp, R. S.: Optimal control of a vehicle suspension incorporating the time delay between front and rear wheel inputs, Vehicle System Dynamics, **17**(6), 1988, 317-336.
6. Kwakernaak, H. and Sivan, R.: Linear Optimal Control Systems, Wiley-Interscience, New York, 1986.
7. Dixon, J. C.: Tires, Suspension and Handling, 2nd ed., SAE, Warrendale, 1996.
8. Sharp, R. S.: The stability and control of motorcycles, Journal of Mechanical Engineering Science, **13**(5), 1971, 316-329.
9. Hahn, B. D.: Essential MATLAB for Scientists and Engineers, Arnold, London, 1997.
10. Valtetsiotis, V.: A discrete time optimal preview control driver model, Cranfield University School of Mechanical Engineering, M. Sc. Thesis, 1999.

LIST OF SYMBOLS

a, b	car dimensions, Fig. 2
h	number of preview points ahead of car
i	summation counter
k	time counter
r	car yaw rate
u	car forward velocity
v	car lateral velocity
y	absolute lateral displacement, Fig. 1
A	state matrix
B	input distribution vector
C	output vector
C	tyre cornering stiffness
D	road shift operator matrix
E	road input description vector

G	handwheel to roadwheel steering ratio
I_z	car yaw inertia
J	cost function
$\mathbf{K}, \mathbf{K}_1, \mathbf{K}_2$	gain vectors, $\mathbf{K}=[\mathbf{K}_1; \mathbf{K}_2]$
M	car mass
$\mathbf{Q}$	weighting matrix
$\mathbf{R}_1$, R_2	weighting matrix and scalar in cost function
T	discrete time increment
$\mathbf{X}$, $X_1 \ldots X_4$	state vector and states
$\mathbf{Z}$	extended state vector
α	tyre slip angle
δ_{sw}	steering wheel angle
ψ	car absolute attitude angle, Fig. 1

subscripts: d, discrete time; f, front; r, rear, road;

Vehicle System Dynamics Supplement 35 (2001), pp. 119-134

Sensitivity To Mass Variations Of The Fastest Possible Lap Of A Formula One Car

D. CASANOVA*, R. S. SHARP** and P. SYMONDS*

SUMMARY

The paper describes methods for setting up and solving theoretically the minimum time lap problem of contemporary Formula One cars. Emphasis is placed on computational efficiency and robustness of the calculations. The problem is treated as one of non-linear programming. Car and circuit models are set up and novel techniques are used to achieve effective optimisation in reasonable computer time. The vehicle representation includes seven degrees of freedom and realistic aerodynamic and tyre force systems. Results relating to the influence of total car mass on the minimum time achievable on two current Formula One circuits are included, to illustrate the kind of information which can be gained from the optimisation. Results concern the range 600-700kg, which spans the practically interesting circumstances. Results are shown to be repeatable from different starting conditions, precise and in general agreement with knowledge from testing.

1. INTRODUCTION

Vehicle modelling is a well established subject and current models allow the representation of the vehicle transient behaviour with high accuracy. The natural extension of simulation is optimisation. The ideal development is to have a formal procedure to evaluate the optimal control strategy and the optimal combination of the design parameters in order to achieve a certain goal.

For a circuit racing car the goal is simply to traverse laps of the circuit in minimum time. Mathematical models have been developed in order to predict the maximum performance as a function of vehicle parameter changes. The most common representation of the performance envelope of a race car is the "*g/g/speed*" surface, which is the envelope of a set of "*g/g*" planar diagrams representing each of the maximum longitudinal and lateral accelerations in steady-state conditions at a certain velocity [1]. The basic approach to lap time simulation assumes that the racing line is known by some means. Then the path round the circuit is divided into arbitrarily short segments with constant curvature where the vehicle operates at its steady-state performance limits [2]. This approach to lap time simulation is based on the assumption that the race car manoeuvring may be treated as a sequence of steady-state limit conditions. Potentially, this is a strong limitation when studying the response to vehicle design parameter changes in terms of total performance, as many parameters have a significant effect on the vehicle transient behaviour. Furthermore, as is well known from practical experience, a race car has to be set up differently for different circuits. When the set-up is altered, the driver has to change his driving strategy in order to bring the car to its new limits. The problem involves coupling between the circuit geometry, the vehicle configuration and the optimal driving strategy. The ideal computer simulation of race car performance must take this coupling into account, as well as including the transient phases.

A novel approach for evaluating the performance of a road vehicle involves the setting up of the minimum time manoeuvring problem as one of Optimal Control. The task is to find the optimal vehicle controls, i.e. the steer angle and the driving/braking torque, which allow the vehicle to perform specified manoeuvres in as short a time as possible while remaining within the road boundaries. Relevant optimal control approaches were reviewed in [3] and [4]. Methods based on the application of Pontryagin's Minimum Principle [5] are called *indirect methods*. They involve the setting up of the adjoint equations, which represent the sensitivities of the objective function with respect to the system states. This results in a non-linear, two point boundary value problem.

Direct methods for the solution of Optimal Control problems are nowadays widely used in practical applications, e.g. in aerospace trajectory optimisation. They are extensively described in the literature of computational optimal control [6]. The basic concept of direct methods is that the continuous control histories are replaced by discrete approximations. It is assumed that the control action can only be adjusted at a number of fixed positions along the trajectory, while the intermediate values are estimated by means of interpolation techniques. The values of the controls at the nodes constitute the set of independent optimisation variables. Objective and constraints are expressed directly as functions of these variables and the problem is solved by using mathematical programming techniques. The intervals over which the controls are determined by interpolation are called "control segments". Direct solution methods do not require the derivation of the adjoint equations, which is of great advantage in practical optimisation problems as it leaves maximum freedom in modelling the system. Also, the discretisation of the control history makes direct methods suitable for those systems whose optimal controls may be discontinuous or rapidly varying, as is the case for the longitudinal control of a racing car. Ideally, the shift between maximum driving and maximum braking will be very rapid [3, 4].

A particular problem which arises when one tries to solve a minimum time problem involving rather long manoeuvres or, indeed, a full lap of a circuit, is the coupling between the early controls and the later parts of the vehicle trajectory. As the early controls influence the entire manoeuvre, they will have a much greater effect on the objective and the constraint functions than the later ones. The last controls applied have little influence on the manoeuvre time or the constraint violations. In the corresponding actual situation, it is as if the driver decided the steer angle control at the beginning of the lap in order to keep the car on the track at the last corner. Sensitivity problems may arise due to this feature. In [3], the *parallel shooting method* for solving the minimum time Optimal Control problem [7] was pioneered. This involves dividing the vehicle trajectory into short, independent segments. These will be referred to subsequently as "trajectory segments". Each control input can only influence the vehicle trajectory within the segment where it occurs, hence decoupling the segment, to some extent, from all the other segments. Clearly, it is necessary to include the requirement for continuity across the segments in the full optimisation problem. The method has proved to be very robust and suitable for the solution of more complex manoeuvres. It was applied in [3] to finding the influence of the vehicle yaw moment of inertia on the maximum possible speed through a double lane change manoeuvre.

In this paper, the direct solution approach of [3] is generalised for the optimisation of entire laps. Issues regarding the discretisation of the problem in order to achieve good robustness and accuracy are discussed. Subsequently, the above framework is applied to investigate the influence of vehicle mass on the performance potential of a Formula One racing car. The actual sensitivity to mass variations is known in general terms from extensive testing and observation, enabling the comparison of the theoretical results with conventional wisdom deriving from testing.

2. THE MINIMUM LAP TIME OPTIMAL CONTROL PROBLEM

Let the differential equations for a general vehicle model, representing the rate of change of its n state variables $\mathbf{x}$ with respect to the time t, be defined as follows:

$$\dot{\mathbf{x}} = \mathbf{a}(\mathbf{x}(t), \mathbf{u}(t), t) \qquad \mathbf{x}(0) = \mathbf{x}_0 \qquad t \in [0, T] \tag{1}$$

Here, $\mathbf{u}(t)$ is a two-element vector function returning the vehicle lateral (i.e. the steer angle) and longitudinal (i.e. the throttle/brake input) controls. Since in the minimum time problem T is not a constant, it is convenient to assume the vehicle travelled distance s, measured on the road centre line, as independent variable for the simulation instead of the time. Therefore, we shall define S_{cf} as the scaling factor relating the increment in time dt to the increment in travelled distance ds:

$$S_{cf}(\mathbf{x}) = \frac{dt}{ds} \tag{2}$$

The expression for the time to distance scaling factor S_{cf} was derived by the present authors in a previous work [8]. Using equation (2), the vehicle model equations become:

$$\begin{aligned} &\frac{d\mathbf{x}}{ds} = S_{cf} \cdot \mathbf{a}(\mathbf{x}(s), \mathbf{u}(s), s) = \bar{\mathbf{a}}(\mathbf{x}(s), \mathbf{u}(s), s) \\ &\mathit{with:} \\ &\mathbf{x}(0) = \mathbf{x}_0 \quad \mathit{and} \quad s \in [0, S] \end{aligned} \tag{3}$$

where S is the length of the road section that the vehicle has to traverse.

The objective function for the Optimal Control problem is the manoeuvre time. From now on, let us adjoin one more state variable x_{n+1} which represents the time elapsed from the beginning of the manoeuvre, and which satisfies:

$$\dot{x}_{n+1} = \frac{dt}{ds} = S_{cf}(\mathbf{x}) \qquad x_{n+1}(0) = 0 \tag{4}$$

Then, the objective function simply reads:

$$F = x_{n+1}(S) \tag{5}$$

The condition for the car to stay on the road is imposed by defining road boundary constraints only where the racing line is expected to lie on the road edges. As is shown in Fig. 1, the racing line tends to use the full road width in specific areas, i.e. before the turn-in phase (1), at the apex of the turn (2) and at the turn exit (3). Along straight track sections, though, it is not necessary to impose any road boundary constraint, as the minimum time problem implies that the vehicle will follow a straight trajectory which minimises the travelled distance. Consider then that m segments where the vehicle lateral position is to be constrained have been identified. Within each segment the distance d of the car's centre of gravity from the road centre line is monitored and the maximum value is used to evaluate the following constraints:

$$c_i = \left(\frac{d_{MAX}}{w_t}\right)^2 - 1 \qquad 0 \le s_i \le S \qquad i = 1..m \tag{6}$$

where, w_t is half the road width. Although this strategy involves some judgement when modelling the circuit, it allows the accurate definition of the road constraints by smooth functions. This is advantageous compared to the use of penalty or barrier functions.

The minimum manoeuvring time Optimal Control problem may now be stated as follows:

$$\begin{aligned}
&\min_{\mathbf{u}} \quad F \\
&\textit{such that} \quad \frac{d\mathbf{x}}{ds} = \bar{\mathbf{a}}(\mathbf{x}(s), \mathbf{u}(s), s) \\
&\qquad\qquad c_i \le 0 \quad i = 1, \ldots, m \\
&\qquad\qquad \mathbf{u}_{\mathrm{L}} \le \mathbf{u}(s) \le \mathbf{u}_{\mathrm{U}} \\
&\textit{for all} \quad s \in [0, S]
\end{aligned} \tag{7}$$

The task is to find the optimal control $\mathbf{u}(s)$ which minimises the objective function and satisfies the state differential equations, the road boundary constraints and the control bounds associated with finite engine torque etc.

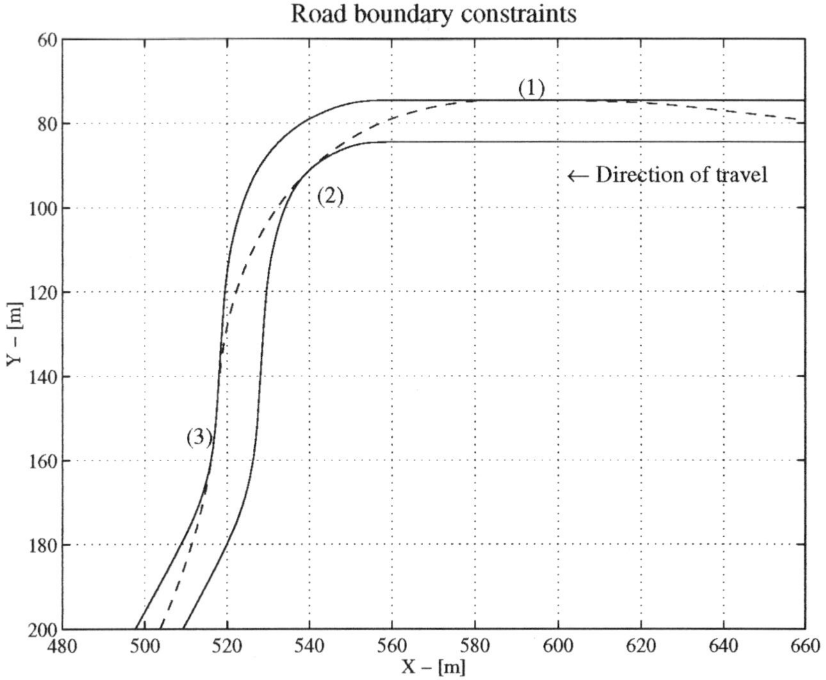

Fig. 1 Road boundary constraints included.

3. THE OPTIMISATION METHOD

A parallel shooting method is used in order to convert the original Optimal Control problem (7) into a *Non-Linear programming problem.* This involves the discretisation of both the state trajectory and the control history over fixed meshes of points, so that a finite number of discrete parameters uniquely identifies the objective and the constraint functions. Typically, a circuit will be divided into about 230 trajectory segments, each containing about 3 to 5 control segments, see section 4. The equations of motion will be integrated over a comparatively very small distance step, so that 3 separate stepwise processes are involved in the optimisation. These three scales are illustrated in Fig. 2.

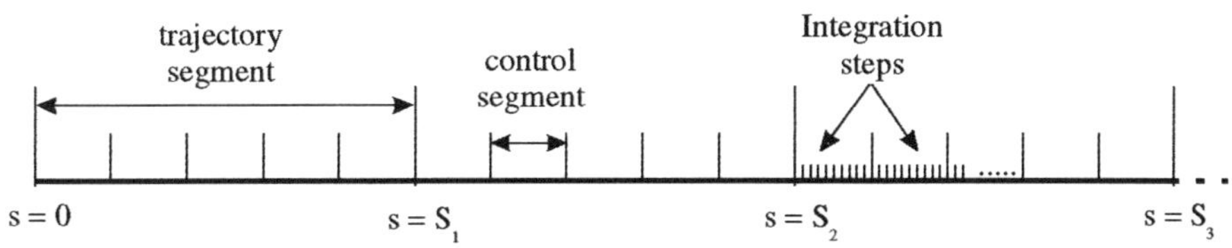

Fig. 2 Illustration of distance scales for trajectory and control segmentation, and integration steps.

Consider the control history first. For simplicity, we shall assume that both the lateral and the longitudinal vehicle controls share the same mesh, although this is not a requirement. Let us define p control intervals as follows:

$$\mathbf{s_p} = \{0 = s_0 < s_1 < s_2 < \cdots < s_{p-1} < s_p = S\} \tag{8}$$

It is assumed that the control input to the vehicle may only be varied at the nodes of the mesh, while the intermediate values are estimated by means of linear interpolation:

$$\mathbf{u}(s) = \text{lin\_interp}(\mathbf{s_p}, \mathbf{u_p}, s) \quad \Rightarrow \quad \mathbf{u} = \mathbf{u}(\mathbf{u_p}) \tag{9}$$

Here, $\mathbf{u_p}$ is the vector of independent control parameters which, for the case of the vehicle that has two control inputs, will have $2p$ elements.

Next, consider the division of the state trajectory into q segments by defining a second mesh of points as follows:

$$\mathbf{s_q} = \{0 = s_0 < s_1 < s_2 < \cdots < s_{q-1} < s_q = S\} \tag{10}$$

To indicate the values of the state variables at the ith node, we shall use the following notation:

$$\mathbf{x}_i = \mathbf{x}(s_i) \qquad i = 0, \ldots, q \tag{11}$$

The values of the state variables at the beginning of the segments constitute the vector of the independent state parameters:

$$\mathbf{x_q} = \{\mathbf{x}_0, \mathbf{x}_1, \mathbf{x}_2, \ldots, \mathbf{x}_{q-1}\} \tag{12}$$

If there are n state variables, the vector $\mathbf{x_q}$ will have $n \times q$ elements. Finally, we shall define the vector $\mathbf{y}$ for all the independent optimisation variables as:

$$\mathbf{y} = \{\mathbf{u_p}, \mathbf{x_q}\} \tag{13}$$

The trajectory within each segment is evaluated by solving the following initial value problem with a suitable numerical integration method, e.g. Runge-Kutta:

$$\begin{aligned} &\frac{d\mathbf{x}}{ds} = \bar{\mathbf{a}}(\mathbf{x}(s), \mathbf{u}(s), s) \\ &\textit{with} \quad \mathbf{x}(s_{i-1}) = \mathbf{x}_{i-1} \quad s \in [s_{i-1}, s_i] \\ &\textit{for} \quad i = 1, \ldots, q \end{aligned} \tag{14}$$

The continuity of the state trajectory is imposed by adding sufficient constraints. For each of the interior nodes of the trajectory discretisation mesh, we shall define the vector of *defects* as follows:

$$\mathbf{z}_i = \mathbf{x}(s_i) - \mathbf{x}_i \qquad i = 1,\ldots,q-1 \tag{15}$$

If a closed lap of a circuit is considered, then one extra defect must be defined in order to account for the continuity of the trajectory at the start-finish line:

$$\mathbf{z}_q = \mathbf{x}(s_q) - \mathbf{x}_0 \tag{16}$$

The continuity is therefore established by imposing that all the defects be equal to zero at the solution. This means that, at the solution, the end conditions of one segment match the start conditions of the next one and the end conditions for the whole lap match the start conditions for the next lap.

The state trajectory within each segment is uniquely determined by the independent state parameters and the control input. Therefore, by virtue of (9) and (13), we may write:

$$\mathbf{x} = \mathbf{x}(\mathbf{y}) \tag{17}$$

Hence, the objective and constraint functions may be expressed directly as functions of the independent optimisation variables, and the original Optimal Control problem (7) has been converted into the following Non-Linear Programming problem:

$$\begin{aligned} &\min_{\mathbf{y}} \quad F(\mathbf{y}) \\ &\textit{such that} \qquad \mathbf{z}_i = \mathbf{0} \qquad i = 1,\ldots,q \\ &\qquad\qquad\qquad c_i \le 0 \qquad i = 1,\ldots,m \\ &\qquad\qquad\qquad \mathbf{y}_\mathrm{L} \le \mathbf{y} \le \mathbf{y}_\mathrm{U} \end{aligned} \tag{18}$$

Problem (18) is solved by using the Sequential Quadratic Programming algorithm SNOPT [9]. This is an algorithm suitable for large-scale, non-linear constrained optimisation problems. It requires the objective and constraint function values and their first order derivatives and it uses this information to set up and solve a sequence of Quadratic Programming sub-problems which approximate the original Non-Linear Programming problem. Depending on the discretisation, the number of independent optimisation variables and constraints for the minimum lap time optimisation problem may easily exceed a thousand. However, as a result of the de-coupling between the trajectory segments, the problem derivatives will be very sparse. SNOPT is designed to take advantage of such sparsity and allows the efficient solution of problems with several thousands of independent variables.

4. ISSUES RELATED TO THE PROBLEM DISCRETISATION

The discretisation of the control history is related to the accuracy of the solution. Increasing the number of nodes in the control mesh will, in general, improve the accuracy but will also increase the problem size. It is therefore convenient to specify shorter control segments for track sections where the control input is expected to vary more rapidly, e.g. in braking areas, compared to sections where the controls are more constant, e.g. when driving along straight sections.

The trajectory segmentation is very important in order to achieve robust behaviour for the optimisation program. The nature of the minimum time vehicle manoeuvring problem implies that the vehicle has to be driven to the very limit of the tyre contact forces. If the rear tyres saturate before the front ones, the vehicle may well become unstable. As a consequence, a small variation in one of the control parameters may cause the vehicle to spin off the road, with disastrous effects on the simulation. Even in cases in which this does not occur, the sensitivity of the objective function and/or some of the constraints with respect to some of the independent optimisation variables may be much greater than the average. The consequence of such bad scaling of the problem derivatives is slow convergence or even failure to converge. In order to avoid such problems, the lengths of the trajectory segments must then be kept sufficiently short so that, according to the vehicle dynamic properties, *a spin cannot occur within the distance of the segment*.

The optimal problem discretisation can only be achieved by doing several trials, progressively refining the mesh in those areas of the circuit where the sensitivity appears to be critical. It is also important to try very different vehicle set-ups, so that robust behaviour is achieved for all anticipated vehicle configurations. For the optimisation problems presented in this paper, the maximum length of a trajectory segment was set to 120m, corresponding to straight track sections, between 20 to 30m for medium and fast bends and as short as 8m for very slow corners. The lengths of the control segments were set from a maximum of 20m (e.g. for straight sections) to a minimum of 4m (e.g. very tight corners).

5. CIRCUIT AND VEHICLE MODELS

The model of a circuit is essentially described by the following (non-independent) parameters, Fig. 3:

- The co-ordinates of its centre line in a reference axes system fixed in space, x_t, y_t;
- The curvature of the centre line, k_t, or its local radius r_t;
- The tangent angle of the centre line, ψ_t;
- The road width, w_t.

These data are expressed as functions of the independent path co-ordinate s, i.e. the distance travelled along the road centre line from the start-finish line.

Complementary information may be supplied, e.g. road elevation, road camber angle, variations in the friction coefficient of the road surface, etc. The track data are supplied to the vehicle model as look-up tables.

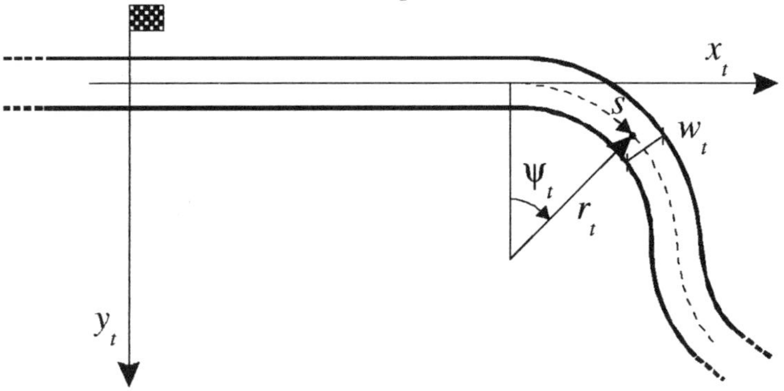

Fig. 3 Description of racing circuit.

The vehicle model that is used in this work for the mass sensitivity studies is similar to that implemented by the present authors in a previous work, with only minor changes. A brief outline is given here, while for the detailed description the reader is referred to [3].

The vehicle is represented with a seven degrees of freedom model. The chassis is treated as a rigid body with three degrees of freedom, the yaw angle and the lateral and longitudinal displacements. The wheels are rigidly attached to the body, allowing only the rotation of each wheel relative to the vehicle chassis about its spin axis. This adds four degrees of freedom to the system. Although the suspensions are not included in the modelling, the roll axis position, the roll stiffness distribution, the mass centre height and the track width are specified. These parameters are sufficient to derive approximate expressions for the lateral and longitudinal load transfers corresponding to a constant acceleration of the mass centre [10].

A simple representation of the aerodynamic forces is employed by assuming constant drag and lift coefficients. The aerodynamic drag is applied at the height of the vehicle centre of gravity. The centre of application of the aerodynamic lift is the same for all speeds and is determined by specifying the down force distribution between the front and the rear axles.

The tyre lateral and longitudinal forces are introduced using the Magic Formula Tyre Model which features the use of weighting functions to account for combined slip conditions [11]. The tyre slip quantities are evaluated with the assumption of small angles (i.e. for a generic angle α it is assumed that $\cos(\alpha) \cong 1$ and $\sin(\alpha) \cong \alpha$) and they are referred to the centre of the contact area of each wheel. Static wheel camber angle settings are also accounted for. Finally, the wheel vertical loads are evaluated accounting for the vehicle static weight distribution, the aerodynamic down-force distribution and the longitudinal and lateral load transfers when the vehicle is driving, braking and cornering.

The vehicle lateral control variable is the steer angle applied to the front wheels. A parallel steer geometry is considered for the steering system, i.e. the right and left front wheels assume the same steer angle. A single control variable is defined for the longitudinal control. It is assumed that this variable represents the throttle aperture when it is positive or a portion of the maximum braking torque available when it is negative. When driving, firstly the gear ratio is selected based upon the vehicle forward velocity. Then, the engine rotational velocity is computed using the mean rotational velocity of the rear wheels and the total gear ratio. This value is used together with the throttle input in order to read the engine output torque from an experimental engine map. Using the gear ratio, the torque available to the driven wheels is computed. The individual torque applied to each wheel is finally evaluated taking into account the effect of the engine moment of inertia and the torque transfer due to a limited slip differential. The simple model used to compute the torque transfer is based on the Salisbury type differential [1]. In this differential, clutch packs are used to progressively lock the two output shafts together. The axles of the differential pinions rest against wedges which are spread by the input torque, so that they apply a pressure to the clutch packs that is proportional to the input torque. The wedges may have different angles to give different characteristics in driving and in overrun. The amount of torque transfer may be evaluated as a constant portion of the input torque. The direction of the torque transfer depends on the sign of the differential velocity of the rear wheels:

$$\Delta T = C \cdot T_{in} \cdot \mathrm{sign}(\omega_r - \omega_l) \tag{19}$$

The torque transfer gain C assumes each of two different values depending on the sign of the input torque T_{in}.

When braking, the longitudinal control variable simply factors the maximum braking torque available. Then the actual braking torque is shared among the front and rear axles using constant coefficients. The engine brake effect is then added to the rear axle. Finally, the brake torque is split evenly between the front wheels, while the effect of the limited slip differential is accounted for at the rear axle.

The lap time optimisation program requires all this information on the circuit and the car and also on the trajectory and control segmentation and the initial vehicle trajectory and complete set of controls. These data are conveniently decided when modelling the circuit and are joined to the track data file. When the optimisation starts, the data file is read once and all the problem parameters are initialised.

The initial vehicle trajectory and controls are derived through the following procedure. An estimated racing line is drawn by selecting points interactively on the computer map of the circuit and using spline interpolation techniques. The vehicle velocity along the initial trajectory is then specified by using a recording from the on board telemetry data system of a real car. However, since it is important than the initial manoeuvre does not exceed the performance boundary for all anticipated vehicle configurations, it is convenient to rescale the recorded velocity by a factor of about 0.8. The evaluation of the initial controls is then solved as an inverse dynamics problem by using the driver model described in [8].

6. MASS SENSITIVITY RESULTS

The lap time optimisation problem has been solved several times for two circuits, Suzuka and Barcelona, with increasing values of the vehicle mass from 600 kg to 700 kg. For each mass value the optimisation has been repeated four times with different initial values for the independent variables, in order to check the consistency of the "optimal" solution. Table 1 summarises the problem parameters. Table 2 reports the results in terms of lap time. Particularly, the comparison between the best lap time, the worst lap time and the average of the four runs indicates the range of values where the solution may fall.

Fig. 4 summarises the best lap times obtained for the two circuits. The sensitivity of the performance with respect to the vehicle mass is nearly constant for the whole range of values considered. It also turned out to be nearly identical for the two circuits, i.e. 0.0369 s/kg for Suzuka and 0.0361 s/kg for Barcelona. Figs 5 and 6 show the vehicle controls and velocity profiles for Suzuka and Barcelona respectively. The solutions for the lightest and the heaviest vehicles are compared. The heaviest vehicle loses time in the straight sections due to the lower acceleration capability. It also loses speed in the fast corners, as both vehicles have the same downforce but the heaviest has the greatest inertial resistance to changing direction. Finally, Figs 7 and 8 show the racing line for the two circuits for the lightest car configuration.

Table 1. Problem parameters.

	SUZUKA	BARCELONA
Circuit length	5864 m	4728 m
N. of segments	279	242
N. of control nodes	919 (steer), 932 (throttle)	834 (steer), 849 (throttle)
N. of variables	4362	3861
N. of constraints	2725	2353

Table 2. Lap time simulation results.

	SUZUKA			BARCELONA		
Mass	Best	Average	Worst	Best	Average	Worst
[kg]	[s]	[s]	[s]	[s]	[s]	[s]
600	1'18"768	1'18"960	1'19"159	1'34"216	1'34"235	1'34"256
625	1'19"760	1'19"848	1'19"945	1'35"200	1'35"335	1'35"484
650	1'20"714	1'20"792	1'20"871	1'36"145	1'36"226	1'36"305
675	1'21"614	1'21"650	1'21"705	1'36"959	1'37"134	1'37"319
700	1'22"455	1'22"551	1'22"633	1'37"822	1'37"890	1'37"940

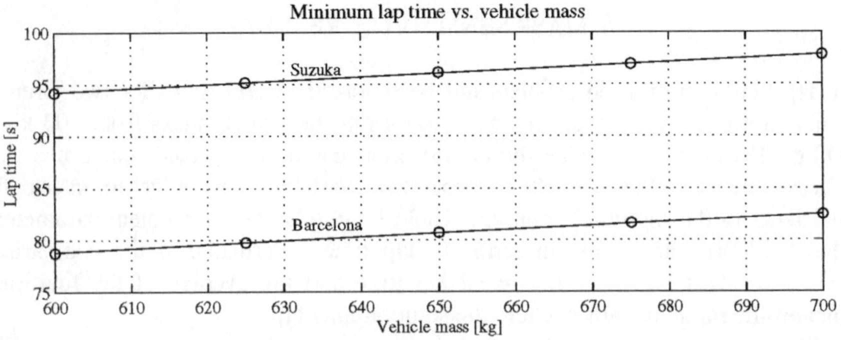

Fig. 4 Minimum lap times as a function of vehicle mass.

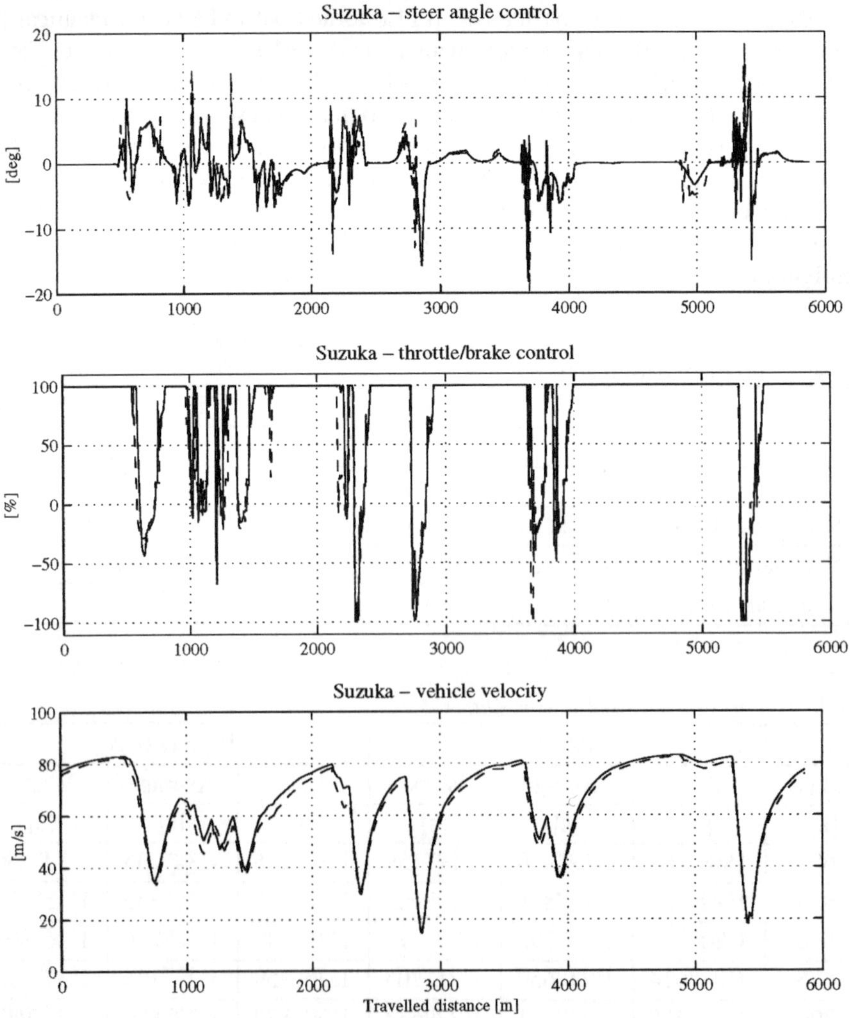

Fig. 5 Vehicle controls and velocity, Suzuka, M = 600 kg (-) and M = 700 kg (---).

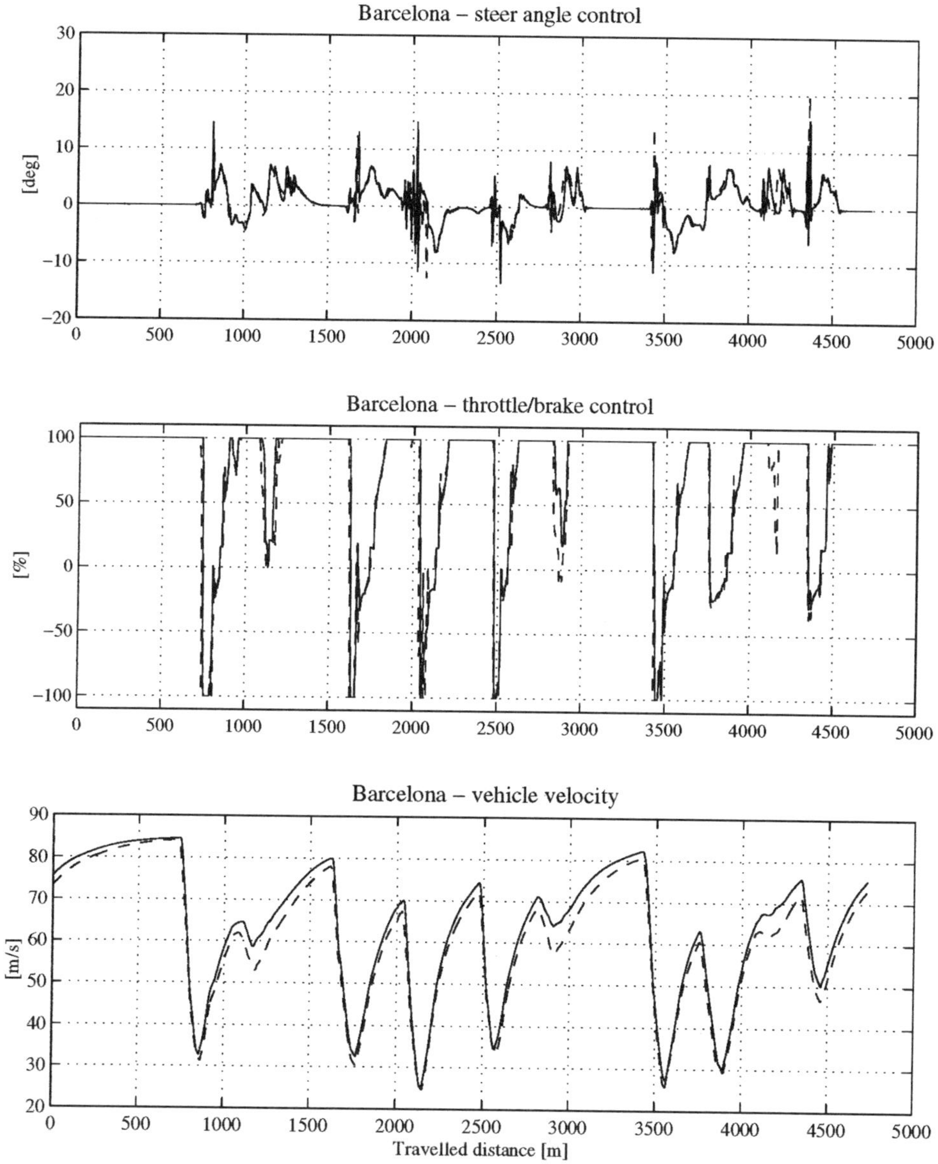

Fig. 6 Vehicle controls and velocity, Barcelona, M = 600 kg (-) and M = 700 kg (---).

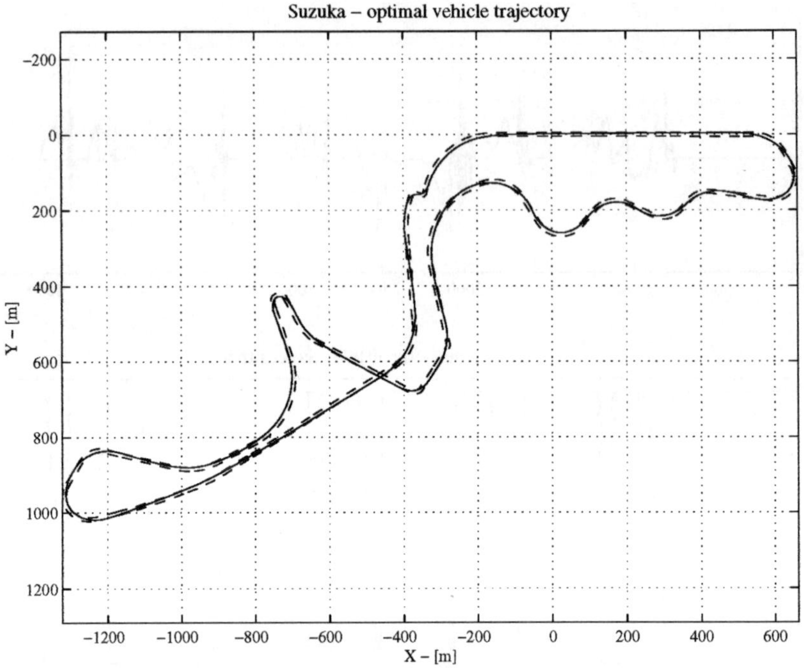

Fig. 7 Racing line, Suzuka circuit, vehicle mass equal to 600 kg.

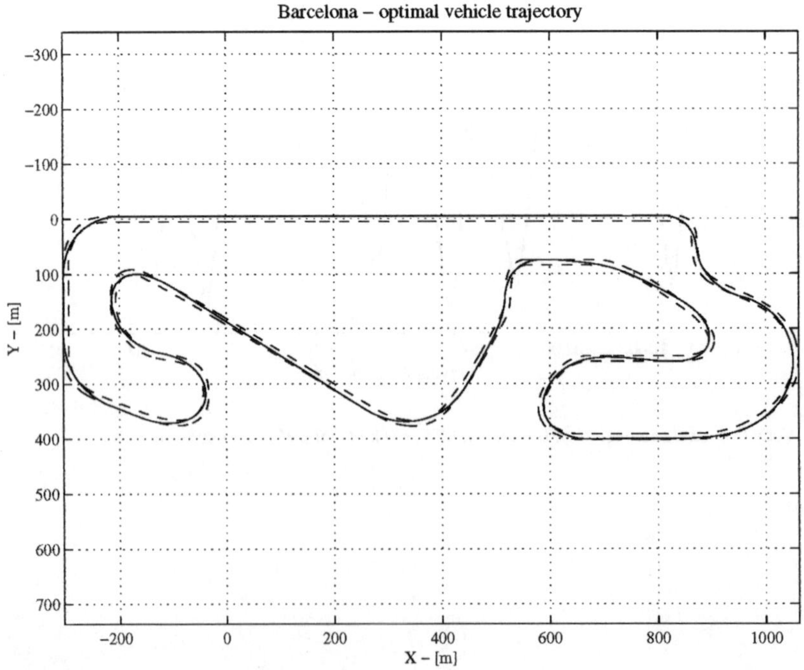

Fig. 8 Racing line, Barcelona circuit, vehicle mass equal to 600 kg.

7. CONCLUSIONS

Mathematical programming techniques have been used for the determination of the optimal lap of a Formula One racing car. The nature of the minimum time vehicle manoeuvre has been exploited in the development of the solution strategy. Excellent robustness and computational efficiency have been achieved when dealing with the large problem of a full lap optimisation. The optimiser finds the racing line and the optimal driving strategy for each case, giving rise to the possibility of learning usefully about many strategic issues in circuit racing.

The method has been validated by studying the influence of the vehicle mass on vehicle performance. For the Suzuka circuit, the sensitivity is 0.0369s/kg/lap, while for the Barcelona circuit, it is 0.0361s/kg/lap. These results are in very good agreement with the sensitivity index that the teams derive from experimentation. The problem is very important in current Formula One racing in relation to the making of strategic decisions for the race.

ACKNOWLEDGEMENTS

The authors should like to thank Prof. P. E. Gill, Dept. of Mathematics, University of California, San Diego, CA, for supplying the optimisation algorithm SNOPT.

REFERENCES

1. Milliken, W.F. and Milliken, D.L.: Race Car Vehicle Dynamics, SAE International, Warrendale, 1995.

2. Griffiths, R.: Minimum Lap Time Simulation of a Racing Car, School of Mechanical Engineering, Cranfield University, MSc thesis, 1992.

3. Casanova, D., Sharp, R. S. and Symonds, P.: Minimum time manoeuvring: The significance of yaw inertia, Vehicle System Dynamics, 34(2), 2000, 77-115.

4. Casanova, D., Sharp, R. S. and Symonds, P.: On minimum time optimisation of Formula One cars: The influence of vehicle mass, Proc. AVEC 2000, Ann Arbor (MI), August, 2000, 585-592.

5. Kirk, D.E.: Optimal Control Theory - An Introduction, Prentice Hall, Englewood Cliffs, New Jersey, 1970.

6. Kraft, D. and Bulirsch, R.: Computational Optimal Control. ISNM, Vol 115, Birkhauser, Basel, Boston, Berlin, 1994.

7. Enright, P. J. and Conway, B. A.: Discrete Approximation to Optimal Trajectories Using Direct Transcription and Nonlinear Programming, AIAA/AHS Astrodynamics Conference, Part 2, Portland, Oregon. Aug. 20-22. 1990.

8. Sharp, R. S., Casanova, D. and Symonds, P.: A mathematical model for driver steering control, with design, tuning and performance results, Vehicle System Dynamics, 33(5), 2000, 289-326.

9. Gill, P. E., Murray, W. and Saunders, M. A.: SNOPT: An SQP algorithm for large-scale constrained optimisation, Numerical Analysis Report 97-2, Department of Mathematics, University of California, San Diego, La Jolla, CA, 1997.

10. Dixon, J. C.: Tires, Suspension and Handling - Second edition, Society of Automotive Engineers, Inc., Warrendale, 1996.

11. Pacejka, H. B. and Besselink, I. J. M.: Magic Formula tyre model with transient properties, in Tyre Models for Vehicle Dynamics Analysis, (F. Böhm and H. – P. Willumeit eds), Suppl. Vehicle System Dynamics 27, 1997, 234-249.

LIST OF SYMBOLS

$\mathbf{a}, \bar{\mathbf{a}}$	vector functions of states, controls and time or distance, representing the vehicle equations, see (1, 3)
$\mathbf{c}$	vector of path constraints, see (6)
d, d_{MAX}	perpendicular distance of car mass centre from track centre-line, see (6)
k_t	track curvature, $1/r_t$
m	number of path constraints
n	order of vehicle model
p	number of control segments
q	number of trajectory segments
r_t	track radius of curvature, $1/k_t$
s	distance along track from start
t	time
$\mathbf{u}$, $\mathbf{u}_L$, $\mathbf{u}_U$	control vector with lower and upper bounds
w_t	half track width
$\mathbf{x}$	state vector
x_t, y_t	Cartesian coordinates of track centre-line
$\mathbf{y}$	$\{\mathbf{u}_p, \mathbf{x}_q\}$ vector of optimisation variables
$\mathbf{z}$	vector of q defects, see (15, 16)
C	differential gear torque transfer gain, see (19)
F	objective function
S	circuit length
S_{cf}	time to distance scaling factor
T	lap time
T_{in}	differential gear input torque
α	slip angle
ψ_t	tangent angle of track at x_t, y_t
ω_r, ω_l	right and left rear wheel spin velocities, see (19)
ΔT	torque transfer across differential gear, see (19)

Vehicle System Dynamics Supplement 35 (2001), pp. 135-148

DYNAMICS OF VEHICLES WITH SEMI-ACTIVE SUSPENSIONS EXHIBITING WHEEL HOP

G. VERROS and S. NATSIAVAS*

SUMMARY

The present study investigates dynamic response of two degree of freedom quarter-car models, subjected to road excitation. In these models, a semi-active control strategy is applied on the suspension dampers, which is based on a switch of their damping coefficient between different values, so that the resulting system approximates the performance of a vehicle involving sky-hook damping. An alternative control strategy is also applied, which is passive and is based on selecting a different damping ratio when the wheel velocity relative to the car body is positive or negative. Both of these control strategies lead to oscillator models with parameter discontinuities. Similar terms are also introduced into the equations of motion by considering wheel hop phenomena, where the vehicle wheel leaves the ground temporarily. Then, application of proper numerical techniques yields results in the form of response diagrams, illustrating the effect of the system parameters on its response and stability properties. Finally, some typical results are also presented for transient road excitation, determined by direct integration of the equations of motion. In all cases, the results are compared to those obtained for vehicles with suspensions including passive shock absorbers. Moreover, special consideration is given to cases where wheel hop is possible to occur.

1. INTRODUCTION

In the design of many new types of automobiles, active and semi-active control strategies based on a proper selection of the vehicle suspension parameters have found extensive application. Moreover, the future trend is to improve and apply such strategies in a rapidly increasing fashion. In order to avoid the complexities and the high cost associated with the construction, instalment and operation of active suspension systems, a lot of research has been devoted in the last three decades to the development of semi-active suspensions, which are easier and less expensive to construct and operate. Starting with the original work of Karnopp et al. [1], several control strategies and mechanical models have been proposed and their effectiveness has been investigated both analytically and experimentally (e.g., [2-8]). The basic

* Department of Mechanical Engineering, Aristotle University, 54006 Thessaloniki, Greece.

idea and objective in all these studies was to improve the vehicle ride and handling performance by using suspension dampers with variable characteristics, in place of active force generators.

Despite the practical advantages introduced in the behavior of vehicles by using semi-active suspensions, the research studies oriented towards developing a better understanding of their performance run into severe difficulties, since the resulting mechanical models are usually strongly nonlinear. This is so because these models involve parameter discontinuities related to the switch in the damping coefficient of the suspension damper. In addition, the response is affected significantly by other types of nonlinearities. Among them, the nonlinearity associated with wheel hop phenomena [9], where a wheel leaves the road temporarily, plays a dominant role on the vehicle ride and handling performance. Typically, proper evaluation of this performance is based on examination of frequency spectra for three types of response quantities, within the range of 0-20 Hz [9, 10]. Namely, the maximum absolute acceleration of the vehicle body, which is related to passenger comfort and isolation, the suspension travel, related to wheel space dimensions and the wheel traction, related to vehicle handling and safety.

The main objective of this study is to investigate some fundamental issues related to the dynamics of semi-actively controlled vehicles, by employing a classical two degree of freedom quarter-car model, subjected to road excitation. The control strategy applied on the suspension absorber is based on a choice between different values of its damping coefficient, so that the system approaches a desirable state of sky-hook damping. Special consideration is given to a careful examination of the wheel hop effects, while attention is focused on the vehicle ride performance. The resulting equations of motion are strongly nonlinear and their solution is obtained by applying special numerical methodologies. In all cases, the ride response of semi-active vehicles is compared to that exhibited by vehicles with conventional passive bilinear or linear suspension dampers.

The organization of the present paper is as follows. First, the mechanical models examined are presented in the following section, together with the control strategy imposed on selecting the equivalent suspension damping coefficient. Then, the methodology employed in detecting and incorporating in the analysis possible loss of contact between the wheel and the ground is presented in the third section. Next, typical results are presented in the form of response diagrams, illustrating the effect of the damping parameters on the response. These results are obtained by application of special numerical methodologies and provide useful information in evaluating the vehicle ride performance and in detecting frequency ranges characterised by complex dynamic behavior. Finally, numerical results are obtained for transient road excitation, by direct integration of the equations of motion. These results are compared to those obtained for the conventional suspension passive shock absorbers. In addition, special consideration is given to cases of wheel hop phenomena.

2. VEHICLE MODELS - CONTROL STRATEGY

The mechanical system examined in the present study is represented by the classical quarter-car models, shown in Figure 1. In particular, if the coordinates x_1 and x_2

represent the absolute displacement of the wheel and the vehicle body, respectively, then the equations of motion of the sky-hook model shown by Figure 1a can be put in the following matrix form

$$M\underline{\ddot{x}} + C\underline{\dot{x}} + K\underline{x} = \underline{f}(t)\,, \qquad (1)$$

where

$$M = \begin{bmatrix} m_1 & 0 \\ 0 & m_2 \end{bmatrix}, \quad C = \begin{bmatrix} c_1 + c_2 + c_g & -c_2 \\ -c_2 & c_2 + c_s \end{bmatrix} \quad \text{and} \quad K = \begin{bmatrix} k_1 + k_2 & -k_2 \\ -k_2 & k_2 \end{bmatrix}$$

are the mass, damping and stiffness matrix of the system, respectively, while the vector $\underline{f}(t)$ represents the external forcing. In the present study, attention is focused on ride vehicle behavior and the external excitation is assumed to result from road irregularities. Therefore, the forcing vector appears in the form

$$\underline{f}(t) = \begin{pmatrix} 1 \\ 0 \end{pmatrix} [k_1 s(t) + (c_1 + c_g)\dot{s}(t)]\,.$$

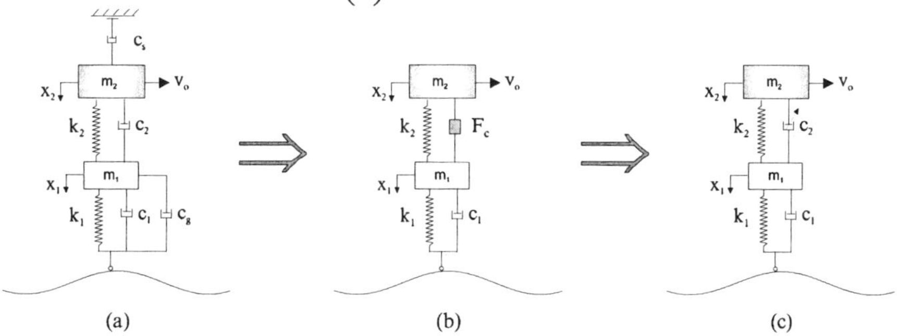

Fig. 1. Vehicle models: (a) sky-hook model; (b) active model; (c) semi-active model.

In the special cases where the vehicle is travelling with a constant horizontal velocity v_0 and the road profile is harmonic, with form $s = s_0 \sin(2\pi z/\lambda + \varphi)$, the forcing vector can be written in the form

$$\underline{f}(t) = \underline{f}_c \cos\Omega t + \underline{f}_s \sin\Omega t\,,$$

with forcing frequency $\Omega = 2\pi v_0/\lambda$ and

$$\underline{f}_c = s_0 \begin{pmatrix} 1 \\ 0 \end{pmatrix} [k_1 \sin\varphi + \Omega(c_1 + c_g)\cos\varphi]\,, \quad \underline{f}_s = s_0 \begin{pmatrix} 1 \\ 0 \end{pmatrix} [k_1 \cos\varphi - \Omega(c_1 + c_g)\sin\varphi]\,.$$

In order for a ground vehicle to achieve the dynamic performance of the sky-hook model shown in Figure 1a, active control must be provided through an actuator, located between the sprung and the unsprung masses, as shown in Figure 1b. However, since active control is associated with high costs and considerable energy consumption, an alternative design methodology is adopted, instead, based on a semi-active logic. Namely, the semi-active controller involves sensors and actuators that can change the suspension damping coefficient, as indicated in Figure 1c. In this way, this control type requires much less installation and operational costs and at the same time maintains the essential advantages obtained through the application of an active force generator.

Among the various control strategies that are currently available, the logic adopted in the present study is based on an optimum strategy applied in earlier studies (e.g., [5, 7]). First, the controller force in the active model of Figure 1b is selected in the form

$$F_c = c_s\dot{x}_2 + c_2(\dot{x}_2 - \dot{x}_1) - c_g(\dot{x}_1 - \dot{s})\,. \qquad (2)$$

Then, if the model shown in Figure 1c is to exhibit similar response with that of the model presented in Figure 1b, its damping coefficient $\hat{c}_2$ must be monitored so that

$$c_s\dot{x}_2 + c_2(\dot{x}_2 - \dot{x}_1) - c_g(\dot{x}_1 - \dot{s}) = \hat{c}_2(\dot{x}_2 - \dot{x}_1)\,.$$

However, instead of continuously changing the value of $\hat{c}_2$, as suggested by the last formula, a more economical and easier to apply strategy is followed. This alternative strategy is based on a semi-active control logic and is realized by a switchable damper, which can provide a variable damping coefficient $\hat{c}_2$ with minimum and maximum value equal to $c_{2\min}$ and $c_{2\max}$, respectively. Namely, the value of $\hat{c}_2$ is selected so that

$$\hat{c}_2 = \begin{cases} c_{2\max}, & if \quad F_c > c_{2\max}(\dot{x}_2 - \dot{x}_1) \\ \dfrac{F_c}{\dot{x}_2 - \dot{x}_1}, & if \quad c_{2\min}(\dot{x}_2 - \dot{x}_1) < F_c \le c_{2\max}(\dot{x}_2 - \dot{x}_1) \\ c_{2\min}, & if \quad F_c \le c_{2\min}(\dot{x}_2 - \dot{x}_1) \end{cases}$$

Finally, vehicle models with passive suspensions are also considered. More specifically, besides the classical linear models (where $\hat{c}_2 = c_{2\min} = c_{2\max}$), another commonly employed passive control strategy is based on switching the value of c_2 between $c_{2\min}$ and $c_{2\max}$, depending on the sign of the relative velocity of the sprung mass to that of the unsprung mass [12, 13].

3. VEHICLE MODELS INVOLVING WHEEL HOP

The derivation of the equations of motion for the mechanical models examined in the previous section was based on the assumption that the vehicle wheel remains always in contact with the ground. However, under certain loading conditions the vehicle wheel may separate from the ground, leading to what is known in the literature as wheel hop [9]. Appearance of such phenomena imply considerable changes in the vehicle response. On the other hand, the resulting vehicle model is more difficult to analyze because the loss of contact introduces strong nonlinearities in the equations of motion, even for vehicle models with constant technical parameters.

For an effective study of the wheel hop phenomenon, it is useful to first introduce the new kinematic variable

$$X_1(t) = x_1(t) + x_{1st} - s(t)\,,$$

with

$$x_{1st} = (m_1 + m_2)g/k_1\,.$$

Then, it is obvious that the condition

$$X_1 < 0 \qquad (5)$$

implies that the wheel is not in contact with the ground. On the other hand, the condition

$$X_1 \geq 0$$

is not sufficient to ensure contact of the wheel with the ground. In this case, the contact condition is judged by also considering the sign of the quantity

$$F_{10} = k_1 X_1 + (c_1 + c_g)\dot{X}_1,$$

which represents the force developed between the ground and the wheel. Therefore, the condition for contact requires that

$$F_{10} \geq 0 \quad \Rightarrow \quad k_1 X_1 \geq -(c_1 + c_g)\dot{X}_1 . \tag{6}$$

For analytical convenience, the results expressed by conditions (5) and (6) can be presented in the $(X_1, \dot{X}_1)$ plane [17]. For instance, based on the above analysis, Figure 2 presents the areas of the $(X_1, \dot{X}_1)$ plane where the vehicle wheel is in contact or looses contact with the ground.

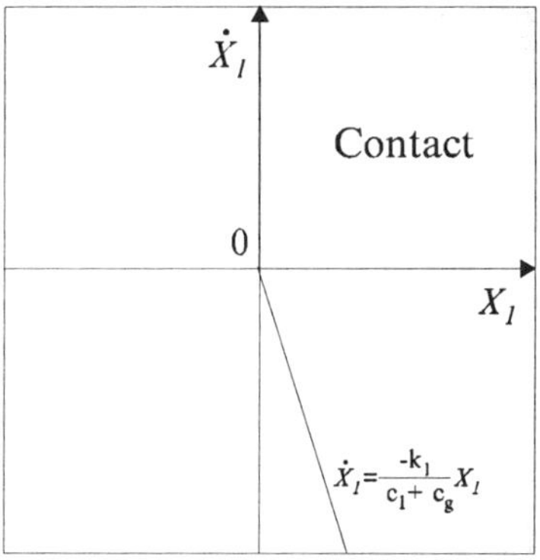

Fig. 2. Areas of the $(X_1, \dot{X}_1)$ plane, indicating contact or loss of contact of the wheel with the ground.

For the time intervals where the vehicle wheel is in contact with the ground, the equations of motion appear in the form (1), presented in the previous section. On the other hand, during the time intervals where the wheel is not in contact with the ground, the mass matrix retains its original form, while the damping matrix, the stiffness matrix and the external forcing vector appear in the form

$$C = \begin{bmatrix} c_2 & -c_2 \\ -c_2 & c_2 \end{bmatrix}, \quad K = \begin{bmatrix} k_2 & -k_2 \\ -k_2 & k_2 \end{bmatrix} \quad \text{and} \quad \underline{f}(t) = \underline{0},$$

respectively.

4. PARAMETRIC STUDY

In this section, typical numerical results are presented for example vehicle models, with nominal parameter values $m_1 = 60kg$, $m_2 = 375kg$, $k_1 = 200kN/m$,

$k_2 = 15 kN/m$, $c_1 = 7 Ns/m$ and $c_2 = 1425 Ns/m$. This set of parameters is first completed by selecting the damping coefficients c_s and c_g through an optimization process. Namely, the values of these parameters are chosen so that they minimize the cost function

$$J(c_s, c_g) = \frac{1}{2T(m_2 g)^2} \int_0^T [(m_2 \ddot{x}_2)^2 + F_w^2(t)] dt \,,$$

at every value of the forcing frequency $\Omega = 2\pi/T$. The term F_w represents the force developed between the ground and the wheel and the calculations were run at steady state conditions, while the maximum value of the damping coefficients was set to $50 kNs/m$. Figure 3a presents the optimum values obtained for the damping coefficients c_s and c_g as a function of the forcing frequency Ω, within the frequency range of interest. The results indicate that these values are strong functions of the forcing frequency Ω. Consequently, in order to make a reasonable selection, the corresponding cost function is also shown in Figure 3b as a function of Ω. From the last diagram it appears that the cost function presents a global maximum at about $\Omega = 12 Hz$. Based on these results, the damping values chosen for all the subsequent calculations were $c_s = 50 kNs/m$ and $c_g = 4 kNs/m$.

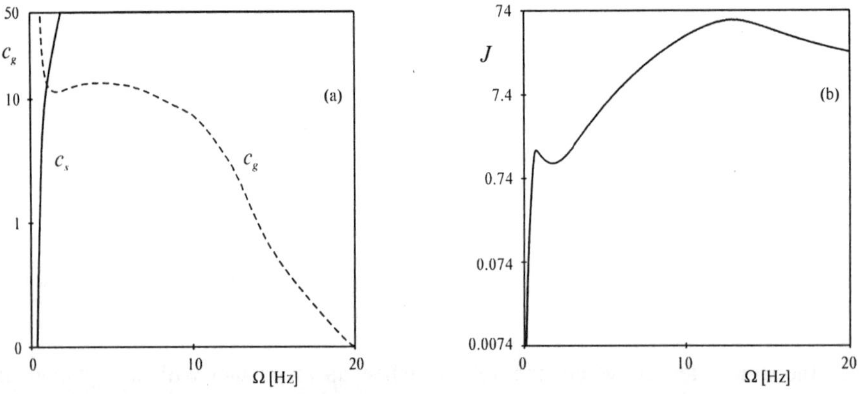

Fig. 3. Selection of optimum damping coefficients: (a) c_s and c_g versus forcing frequency Ω; (b) cost function $J(c_s, c_g)$ versus Ω.

Next, numerical results are presented in the form of response diagrams. Initially, Figure 4 shows typical response diagrams for periodic motions, obtained for several versions of the sky-hook vehicle model shown in Figure 1a, by ignoring wheel hop. All these models are linear and Figures 4a and 4b present the response spectra for the absolute acceleration of the vehicle body and the force developed between the wheel and the ground, respectively. The vehicle is assumed to run over a harmonic road, with $s_0 = 0.02 m$ and $\varphi = 0$. In both figures, significant differences are observed to exist between the cases with $c_s = c_g = 0$ (corresponding to the conventional passive quarter car model) and $c_s c_g \neq 0$. Also, the effect of the c_s value is found to be more important in the low frequency ranges, while the effect of the value of c_g is more evident in higher frequencies, since they mostly affect the motion of the vehicle body in the first and second mode shape, respectively. Finally, the generalization of the classical Karnopp's model (corresponding to $c_2 = c_g = 0$) presents the best behavior near the first resonance, but is characterised with the worse behavior in the vicinity of the second resonance.

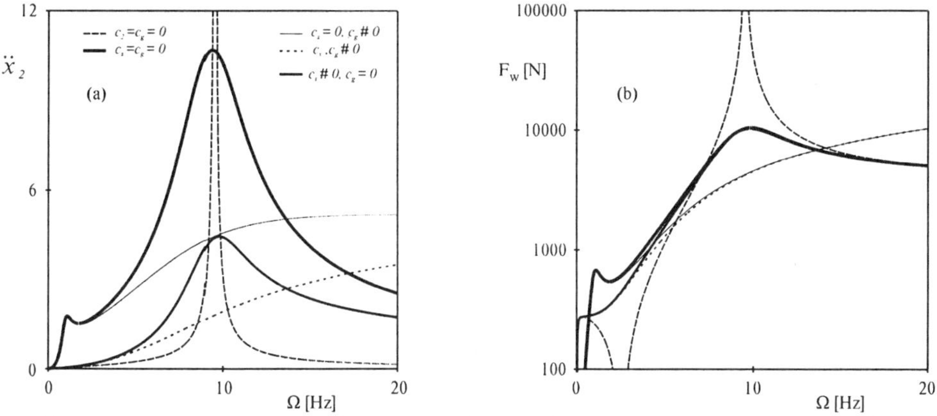

Fig. 4. Response diagrams for the sky-hook vehicle model: (a) absolute vehicle acceleration $\ddot{x}_2$ and (b) wheel force F_w.

Next, similar results are presented for nonlinear vehicle models. Since the model nonlinearities are due to switches between constant values of the suspension damping coefficient, appropriate methodologies can be applied for obtaining exact periodic motions resulting from roads with periodic profile, like those presented in [11] and [14]. However, when the wheel hop possibility is included, these motions may involve many solution intervals, complicating the solution process. For this reason, the steady state response and stability of the models examined were determined by application of suitable numerical methodologies, instead (e.g., [15, 16]). In this way, complete branches of periodic steady state responses can be determined, together with their stability properties, providing useful information for a systematic evaluation of the vehicle ride performance.

The following sequence of response diagrams compares results obtained for vehicle models including or excluding the possibility of wheel hop. Due to the asymmetry induced in the response by the wheel hop, both the maximum and the minimum values of the vehicle acceleration and the wheel force are recorded in these diagrams. First, Figure 5 presents results for the semi-active model (Figure 1c). As indicated by the thinner lines, the wheel hop phenomena start appearing in the vicinity of the second linear natural frequency and continue to exist up to the end of the frequency range examined. Moreover, apart from a relatively narrow frequency interval at the frequency where wheel hop initiates, the response of the model with wheel hop is found to be smaller than that predicted by the model without wheel hop. In addition, all the periodic responses shown are found to always be stable.

Quite similar observations can be made for a vehicle model with passive bilinear shock absorbers, based on the results presented in Figure 6. These absorbers are designed so that they exhibit a dual-rate behavior with $c_{2\min} = 475\,Ns/m$ and

$c_{2\max} = 1425\,Ns/m$, while $c_s = c_g = 0$. In this case, however, branches of unstable periodic motions (represented by broken lines) are obtained for the model with wheel hop, within two specific frequency ranges. The first of these branches appears in the vicinity of the second linear natural frequency through saddle-node bifurcations at its ends, while the other appears at higher forcing frequencies (namely within the frequency interval $\Omega = 14.3 - 17.6Hz$), through a period doubling at both of its ends [16]. As shown in the diagrams, the unstable branch of $n = 1$ motions generated by the period doubling bifurcations gives rise to an $n = 2$ branch of stable motions.

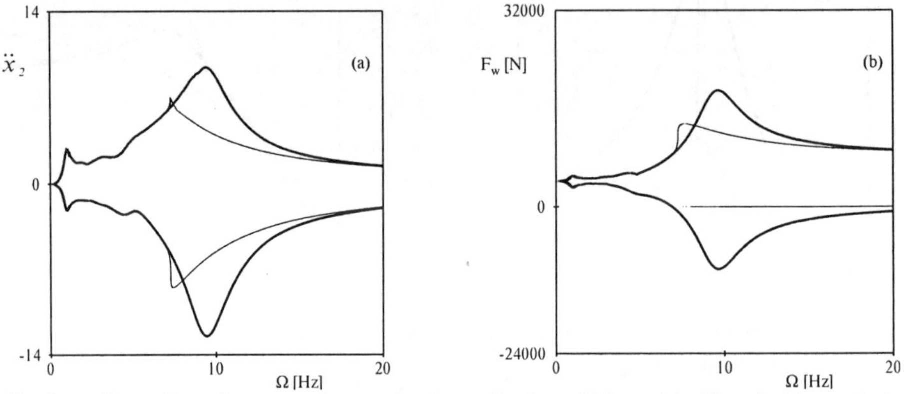

Fig. 5. Comparison of response diagrams for the semi-active vehicle model with and without wheel hop (indicated by the thick and the thin lines, respectively): (a) absolute vehicle acceleration $\ddot{x}_2$, (b) wheel force F_w.

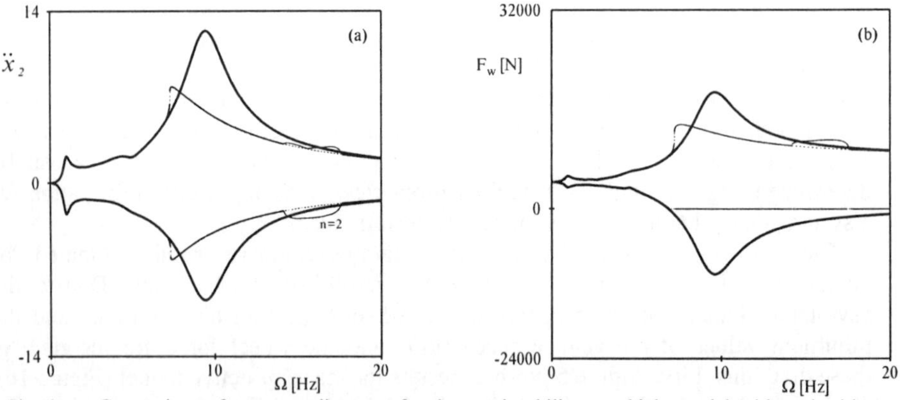

Fig. 6. Comparison of response diagrams for the passive bilinear vehicle model with and without wheel hop (indicated by the thick and the thin lines, respectively): (a) absolute vehicle acceleration $\ddot{x}_2$, (b) wheel force F_w.

Finally, Figure 7 presents similar diagrams for a vehicle possessing a linear suspension absorber, with $c_s = c_g = 0$ and constant c_2. In particular, when $c_2 = 1425\,Ns/m$, the $n = 1$ periodic response is found to always be stable, as shown in Figures 7a and 7b. However, for $c_2 = 475\,Ns/m$, the $n = 1$ solution branch of the model with wheel hop becomes unstable in several frequency intervals, as shown in Figures 7c and 7d. Again, the first of these intervals appears in the vicinity of the second linear natural frequency through saddle-node bifurcations, while the others appear at higher forcing frequencies. In the latter case, the branch of unstable motions is generated through supercritical period doubling bifurcations at its ends, giving rise to $n = 2$ periodic solution branches. Then, these new branches are also found to loose stability within the $\Omega = 12.57 - 13.31 Hz$ frequency interval, through Hopf bifurcation.

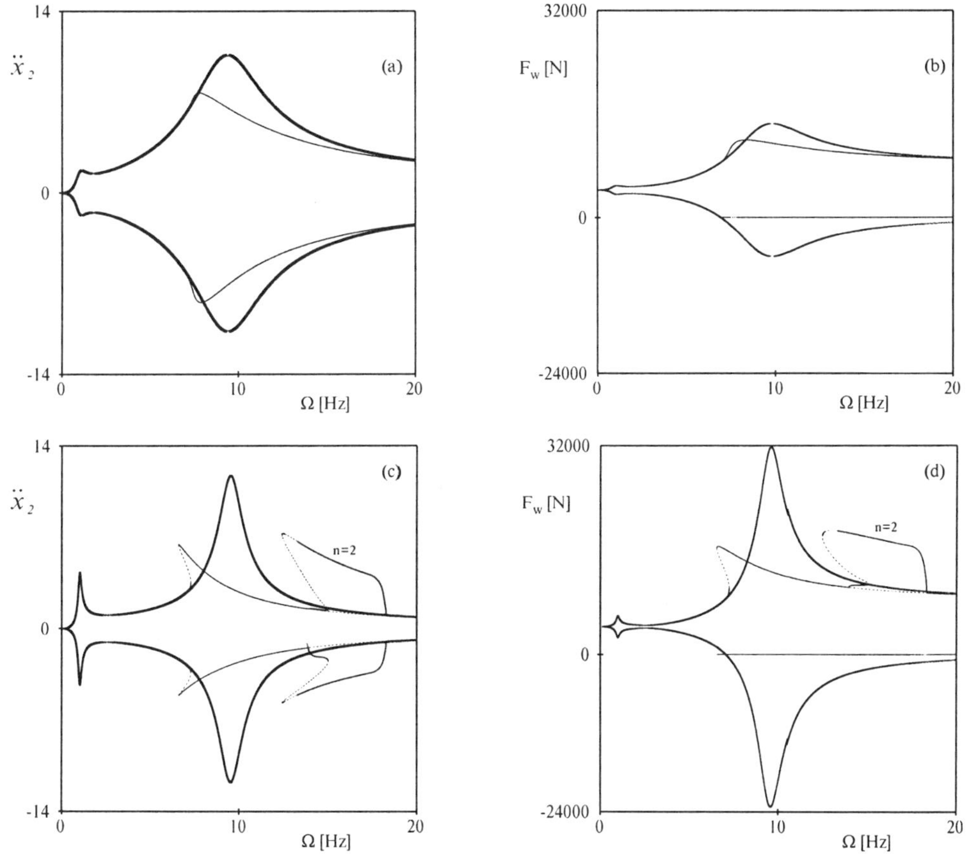

Fig. 7. Comparison of response diagrams for the passive linear vehicle model with and without wheel hop (indicated by the thick and the thin lines, respectively): (a) absolute vehicle acceleration for $c_2 = 1425\, Ns/m$, (b) wheel force for $c_2 = 1425\, Ns/m$, (c) absolute vehicle acceleration for $c_2 = 475\, Ns/m$ and (d) wheel force for $c_2 = 475\, Ns/m$.

Typically, the dynamics exhibited by a system in the neighbourhood of a Hopf bifurcation can be very rich [16]. For the system examined, this is verified by the Poincare sections presented in Figure 8a on the $(x_2, \dot{x}_2)$ plane, at different forcing frequency values. These sections illustrate the transition from $n = 2$ periodic motions to motions with periodically modulated amplitude and back to $n = 2$ periodic motions. In particular, the existence of motions with periodically modulated amplitude is demonstrated further by the history of the corresponding vehicle acceleration at $\Omega = 12.29Hz$, shown in Figure 8b.

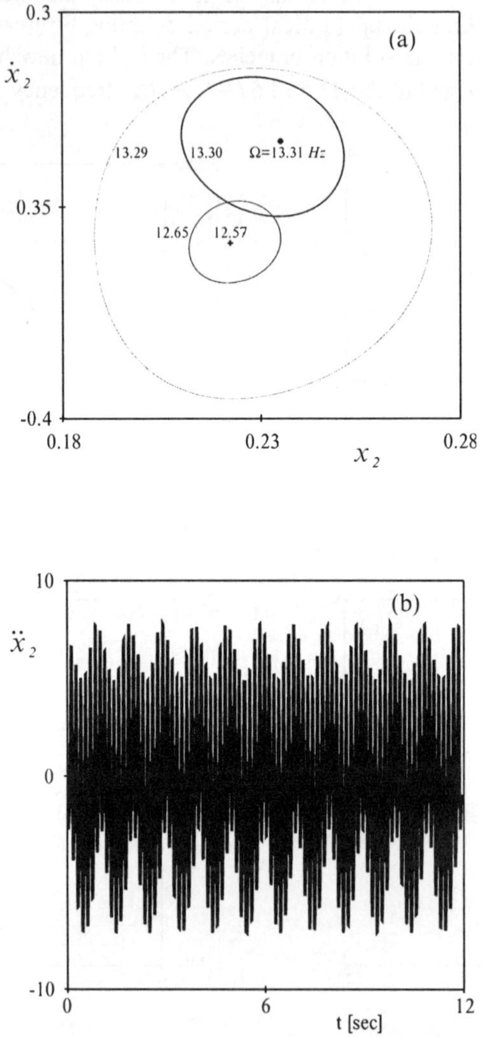

Fig. 8. (a) Poincare sections with the ($x_2, \dot{x}_2$) plane. (b) History of absolute vehicle acceleration at $\Omega = 12.9Hz$.

Obviously, the lower and higher value of the damping coefficient selected for the vehicle with the linear suspension absorber coincide with the minimum and maximum value of the damping coefficient selected for the vehicle with the bilinear suspension absorber. Here, it is mentioned that by increasing the value of c_2 from its minimum to its maximum value, the branch of $n = 2$ periodic motions observed in Figures 7c and 7d separates from the $n = 1$ branch and originally forms an isola, which shrinks gradually and eventually disappears.

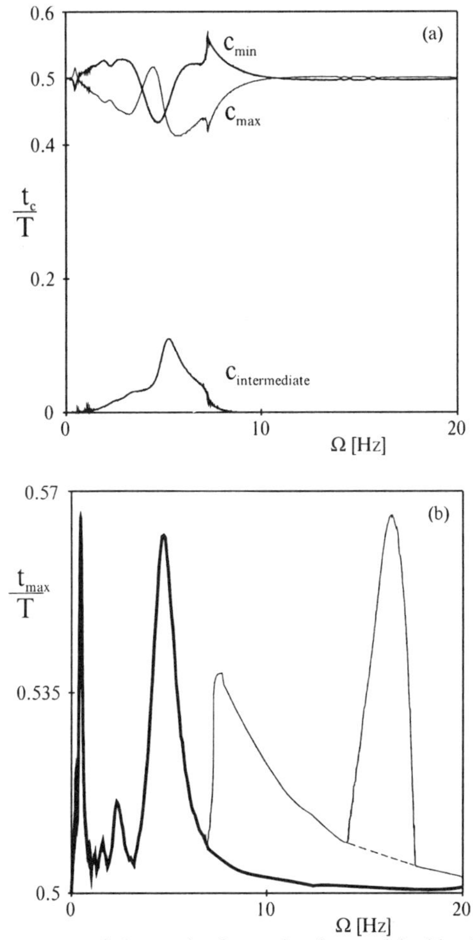

Fig. 9. (a) Time per response period spent by the semi-active model with wheel hop at each of the suspension damping coefficients; (b) Time per response period spent by the passive bilinear model with and without wheel hop (indicated by the thick and the thin lines, respectively) at the maximum value of the suspension damping coefficient.

Finally, Figure 9a presents some information which is useful in assessing the percentage of time spent with the maximum, the minimum and the intermediate suspension damping coefficient within the periodic motion of the model with the semi-active suspension. Clearly, near the frequency of the lowest resonance the system prefers the maximum damping value [18], while an intermediate value is selected mostly within the frequency interval between the two resonances. In the high frequency regime, the system spends almost the same time between the maximum and the minimum damping value.

Similar information is presented in Figure 9b, for the vehicle with the passive bilinear absorber. Again, the thick line shows results for the model where no wheel hop is allowed, while the thin line presents results for the model including wheel hop. These results demonstrate that the appearance of wheel hop is related with a choice of lower suspension damping levels.

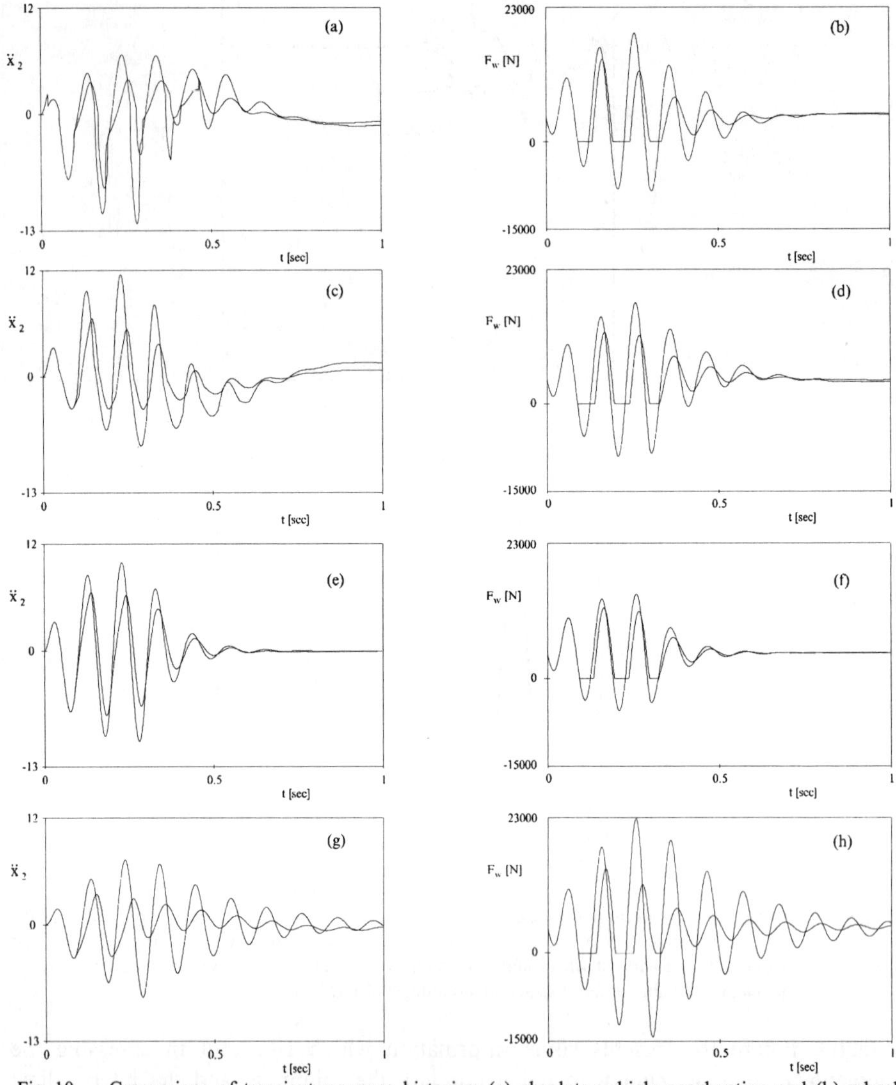

Fig. 10. Comparison of transient response histories: (a) absolute vehicle acceleration and (b) wheel force for the semi-active model; (c) absolute vehicle acceleration and (d) wheel force for the passive bilinear model; (e) absolute vehicle acceleration and (f) wheel force for the passive linear model with $c_2 = 1425\,Ns/m$; (g) absolute vehicle acceleration and (h) wheel force for the passive linear model with $c_2 = 475\,Ns/m$.

5. TRANSIENT ROAD EXCITATION

This section presents numerical results obtained for the quarter-car models examined in the previous section, under transient road excitation. More specifically, these models are assumed to pass over a road presenting an isolated irregularity of the form

$$s(z) = s_0 \sin(2\pi z/\lambda),$$

within the interval $0 \leq z < 3\lambda$. Here, the road wavelength λ and the horizontal vehicle velocity v_0 are chosen so that the resulting forcing frequency is $\Omega = 10Hz$. In all cases, the response is determined by direct integration of the equations of motion.

First, Figures 10a and 10b show the absolute vehicle acceleration and the force developed between the wheel and the ground, respectively, for the model with semi-active suspension. The thicker line represents the results obtained for the model not exhibiting wheel hop, while the thinner line shows the output of the model including wheel hop. Similar results are presented in Figures 10c and 10d for the vehicle with the passive bilinear suspension absorber. Finally, similar diagrams are also presented in Figures 10e through 10h for the model with linear suspension absorbers and values of c_2 coinciding with the maximum and the minimum damping coefficient value chosen for the vehicle with the bilinear suspension absorber. Comparison of the results represented by the thick and the thin lines in Figure 10 indicates that the response of the model including wheel hop is smaller in all the cases examined, in agreement with the results obtained for the harmonic excitation.

6. SYNOPSIS AND CONCLUSIONS

Dynamics of two degree of freedom quarter-car models with semi-active suspensions was investigated. First, the equations of motion of these models were set up for excitation caused by ground irregularities. Then, a control strategy was applied, so that the desired performance of the system is achieved by varying the damping coefficient of the suspension absorber. Finally, the modeling was completed by developing an appropriate methodology, which detects and incorporates wheel hop phenomena. This analysis was then applied and numerical results were obtained, illustrating the effect of the model parameters on its dynamics. Emphasis was put on studying the appearance and effect of wheel hop. In particular, the effect of the forcing frequency and the suspension damping coefficients was examined by first presenting a sequence of response spectra, related to the model ride performance. These spectra were determined through the application of suitable numerical methodologies, which at the same time provided the stability properties of the located periodic motions. In this way, areas of dynamic instabilities were also identified. Through direct integration of the equations of motion at those areas, it was then shown that the models examined may exhibit complex dynamic behavior. In a similar manner, the system dynamics was then investigated further by considering response to transient road excitation. In all cases, attention was focused on comparing the response of the semi-active system with the response obtained from conventional passive linear and bilinear absorbers. Among other things, it was shown that the appearance of wheel hop phenomena may lead to the development of smaller levels of vehicle acceleration and suspension loading.

ACKNOWLEDGEMENT

This research was partially supported by a grant from the General Secretariat of Research and Technology, Greek Ministry of Development, through the PENED99 program.

REFERENCES

1. Karnopp, D., Crosby, M.J. and Harwood, R.A.: Vibration control using semi-active generators. Transactions of ASME, J. of Engineering for Industry, 96 (1974), pp. 619-626.
2. Sharp, R.S. and Hassan, A.S.: The relative performance capabilities of passive, active and semi-active car suspensions. Proceedings of Institution of Mechanical Engineers 20 D3 (1986), pp. 219-228.
3. Palkovics, L. and Venhovens, P.J.Th.: Investigation on stability and possible chaotic motions in the controlled wheel suspension system. Vehicle System Dynamics 21 (1992), pp. 269-296.
4. Lieh, J.: Semiactive damping control of vibrations in automobiles. Transactions of ASME, J. of Vibration and Acoustics 115 (1993), pp. 340-343.
5. Hac, A. and Youn, I.: Optimal design of active and semi-active suspensions including time delays and preview. Transactions of ASME, J. of Vibration and Acoustics 115 (1993), pp. 498-508.
6. Abd El-Tawwab, A.M. and Crolla, D.A.: An experimental and theoretical study of a switchable damper. SAE Paper 960937 (1996), pp. 1177-1190.
7. Valasek, M., Novak, M., Sika, Z. and Vaculin, O.: Extended ground hook - new concept of semi-active control of truck's suspension. Vehicle System Dynamics, 27 (1997), pp. 289-303.
8. Hwang, S.-H., Heo, S.-J., Kim, H.-S. and Lee, K.-I.: Vehicle dynamic analysis and evaluation of continuously controlled semi-active suspensions using hardware-in-the-loop simulation. Vehicle System Dynamics, 27 (1997), pp. 423-434.
9. Gillespie, T.D.: Fundamentals of Vehicle Dynamics. Society of Automotive Engineers, Warrendale, PA, 1992.
10. Dixon, J.C.: Tires, Suspension and Handling. Society of Automotive Engineers, Warrendale, PA, 1996.
11. Verros, G., Natsiavas, S. and Stepan, G.: Control and dynamics of quarter-car models with dual-rate damping. Journal of Vibration and Control 6 (2000), pp. 1045-1063.
12. Wallaschek, J.: Dynamics of non-linear automobile shock-absorbers. International Journal of Non-Linear Mechanics 25 (1990), pp. 299-308.
13. Surace, C., Worden, K. and Tomlinson, G.R.: An improved nonlinear model for an automotive shock absorber. Nonlinear Dynamics 3 (1992), pp. 413-429.
14. Natsiavas, S.: Dynamics of multiple degree of freedom oscillators with colliding components. Journal of Sound and Vibration 165 (1993), pp. 439-453.
15. Doedel, E.: AUTO: Software for Continuation and Bifurcation Problems in Ordinary Differential Equations. California Institute of Technology, Pasadena, California, 1986.
16. Nayfeh, A.H. and Balachandran, B.: Applied Nonlinear Dynamics, Wiley-Interscience, New York, 1995.
17. Leine, R.: Bifurcations in Discontinuous Mechanical Systems of Filippov-Type. PhD Dissertation, Technical University of Eindhoven, Netherlands, 2000.
18. Thomson, W.T.: Theory of Vibration with Applications. Prentice Hall, Englewood Cliffs, New Jersey, 1988.

Vehicle System Dynamics Supplement 35 (2001), pp. 149-164

Lateral and Longitudinal Control of a Vehicle Convoy

AXEL FRITZ *

SUMMARY

Automation of driving requires intelligent vehicle control in longitudinal and lateral direction. For the dynamical analysis of such controls a convoy of two BMW passenger cars is used. To simplify the control problem, in a previous work, the lateral and the longitudinal motion are assumed to be decoupled, and the modeling as well as the controlling of the vehicles in both directions are considered independently. In this paper, a global control concept for the lateral and the longitudinal motion is regarded. Therefore, the convoy is modeled in different complexity for controller design and simulation by the method of multibody systems. On the basis of the more simple model a global control design consisting of a state space transformation and a tracking algorithm is presented. Simulation results for a double lane change with variable speed are shown and the assets and drawbacks of the independent and the global design as well as different control strategies are discussed in detail.

1 INTRODUCTION

Since the various control units in modern vehicles become more and more complex, global networked structures have to replace the isolated units. As an example, in this paper, the lateral and the longitudinal control of a vehicle convoy is regarded. Since the two motions were assumed to be decoupled, in a first project, see [1], only the lateral dynamics and control of the convoy was examined. In that work, the leading vehicle was controlled by a driver and the second vehicle was steered by an actuator to follow autonomously. Considering lateral dynamics only, both vehicles were connected by a rigid tow-bar, see Figure 1, which serves as an ideal longitudinal controller maintaining a constant distance between the vehicles all the time. Then, the longitudinal velocity could be treated as a system parameter instead of a state variable.

In a second project, see [2], an adaptive cruise control (ACC) for the convoy was designed and tested. For the ACC only on board sensors in the following vehicle were used, communication within the convoy or between the controlled

*Institute B of Mechanics, University of Stuttgart, Pfaffenwaldring 9, 70550 Stuttgart, Germany. Tel.: +49-711-6856394. Fax: +49-711-6856400. E-mail: af@mechb.uni-stuttgart.de

Figure 1: Lateral control of the vehicle convoy

vehicle and electronic systems on the roadside was not assumed. A laser scanner replacing the tow-bar was applied for range measurements, see Figure 2.

Figure 2: Longitudinal control of the convoy

In contrast to these two independent control designs, in this paper, a global control design for the lateral and the longitudinal direction is discussed. The paper is organized as follows: In section 2 a simulation model and a control design model for both vehicles are introduced describing the longitudinal and lateral dynamics, respectively. In section 3, first different control strategies are discussed. Then, the global control design is presented and verified by simulation in section 4.

2 VEHICLE MODELS

A simple vehicle model is applied for the global control design to limit the effort while for the controller validation a more complex simulation model is chosen.

2.1 Simulation Model

The simulation model for the lateral and the longitudinal dynamics is based on the method of multibody systems [3]. Each vehicle model is composed of a car body which consists of 9 rigid bodies with 14 degrees of freedom and of a power train model with 4 rigid bodies and 4 degrees of freedom. Additional to the mechanical parts, the combustion process in the engine is regarded as a first order system. For simulation only, the model of the following vehicle is supplemented with two modules representing the actuator dynamics for the steering and the throttle angle while for the leading vehicle a driver model, see [4], is applied.

2.2 Control Design Model

Since the simulation model is too complex to be used for controller design a more simple vehicle model on the basis of the bicycle model of Riekert and Schunck [5] is chosen. Contrary to the bicycle model, in this paper, the longitudinal dynamics is taken into account and only the vertical, the roll and the pitch motion are neglected. The convoy consisting of two of such models is shown in Figure 3.

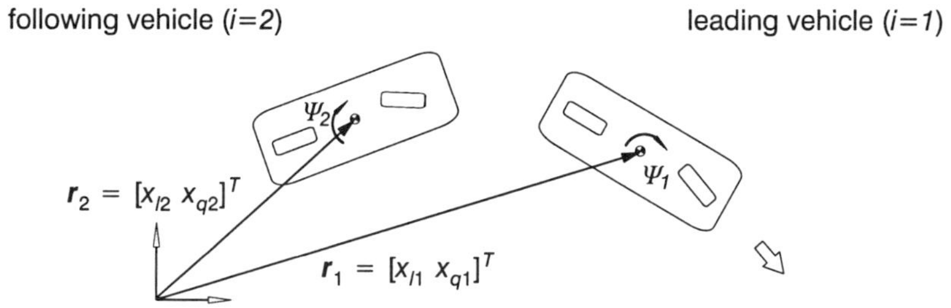

Figure 3: Convoy modeled with two extended Riekert and Schunck models

Both models consist of one rigid body with three degrees of freedom. The generalized coordinates of the vehicle i are the longitudinal displacement x_{li}, the lateral displacement x_{qi} and the yaw angle ψ_i. For the calculation of the lateral tire forces linear relations are considered while the longitudinal slip and the longitudinal tire forces are neglected, respectively. Then, the equations of motion for the vehicle i can be written as

$$\begin{bmatrix} m_q + 2\frac{j_r + j_{red}}{r^2} & \frac{j_r}{r^2}\delta_i & \frac{j_r}{r^2} l_v \delta_i \\ \frac{j_r}{r^2}\delta_i & m_q & m_r(l_v - l_h) \\ \frac{j_r}{r^2} l_v \delta_i & m_r(l_v - l_h) & j_z + m_r(l_v^2 + l_h^2) \end{bmatrix} \begin{bmatrix} a_{li} \\ a_{qi} \\ \ddot{\psi}_i \end{bmatrix}$$

$$= \begin{bmatrix} \frac{k_m}{r} t_{ei} - k_{rv} - k_{rh} - k_c - k_l v_{li}^2 + \frac{l_v \dot{\psi}_i + v_{qi}}{v_{li}} C_v \delta_i \\ C_h \frac{l_h \dot{\psi}_i - v_{qi}}{v_{li}} + C_v(\delta_i - \frac{l_v \dot{\psi}_i + v_{qi}}{v_{li}}) - k_{rv}\delta_i \\ l_v C_v(\delta_i - \frac{l_v \dot{\psi}_i + v_{qi}}{v_{li}}) - l_h C_h \frac{l_h \dot{\psi}_i - v_{qi}}{v_{li}} - k_{rv} l_v \delta_i \end{bmatrix}, \quad (1)$$

where $a_{li} = \dot{v}_{li} - \dot{\psi} v_{qi}$ and $a_{qi} = \dot{v}_{qi} + \dot{\psi} v_{li}$. In the equations (1) m_q denotes the mass of the car body and the four wheels, j_{red} is the moment of inertia of the power train reduced to the rear wheels, j_z is the moment of inertia of the car body related to the vertical direction, m_r is the mass of two wheels, j_r is the moment of inertia of two wheels and r is the wheel radius. The distance from the center of gravity to the front and rear axle are given by l_v and l_h while the front and rear cornering stiffness are expressed by C_v and C_h. The applied forces on the right hand side of Eq. (1) are the rolling, air and climbing drag, $k_r = k_{rv} + k_{rh}, k_l v_{li}^2$ and k_c, where v means front and h rear, respectively. The inputs of the vehicle model are the steering angle δ_i and the delayed engine torque t_{ei} which is multiplied with a coefficient k_m characterizing the power train transmission.

The mechanical model of the vehicle i is enhanced by the combustion process in the engine which is regarded as a first order system with a time constant T_1.

$$T_1 \dot{t}_{ei} + t_{ei} = t_{mi} \tag{2}$$

The engine torque t_{mi} is considered as a linear function of engine speed and throttle position. Neglecting the slip and the elasticity of the power train, the engine speed is proportional to the vehicle speed v_{li}. Then, the engine torque reads as

$$t_{mi} = m_0 + m_1 v_{li} + m_2 \alpha_{thi} \quad , \tag{3}$$

where $m_n, n = 0(1)2$ are constants describing the engine characteristics.

For the control design discussed in the next section only the equations of the following vehicle are considered. Then, the state space vector of the design model corresponds with the state vector of the following vehicle $\boldsymbol{x}_2$ and summarizes the longitudinal and lateral displacements x_{l2} and x_{q2}, the longitudinal and lateral velocities $v_{l2} = \dot{x}_{l2} - \dot{\psi}_2 x_{q2}$ and $v_{q2} = \dot{x}_{q2} + \dot{\psi}_2 x_{l2}$, the yaw angle ψ_2, the yaw rate $\dot{\psi}_2$ as well as the delayed engine torque t_{e2} while the input vector is given by the steering angle δ_2 and the throttle angle α_{th2}. Due to quadratic small terms the coupling term $C_v \delta_2 (l_v \dot{\psi}_2 + v_{q2})/v_{l2}$ in the longitudinal equation of motion is neglected. Furthermore, the wheel mass m_r is set to zero and as a result the elements of the secondary diagonal of the mass matrix vanish.

3 NONLINEAR CONTROL

In this section, first the structure of the closed loop is discussed. Then the different control strategies with measured variables are examined and finally, a global nonlinear control design for the longitudinal and the lateral direction is presented.

3.1 Control Structure

For comparison the independent control of the convoy in longitudinal and in lateral direction is considered too. A linear PDT_1-controller calculates a desired steering angle which is transmitted to the controlled vehicle in a cascade control loop representing the actuator of the lateral control. For the longitudinal control, the model of the following vehicle is supplemented with a nonlinear servo-motor which is realized as a cascade control of the throttle. The servo-motor is used as an actuator transferring the commanded throttle angle from the nonlinear adaptive cruise controller to the vehicle.

Since there is a coupling between the equations of motion in lateral and longitudinal direction, see Eq. (1), here, a global control concept is regarded where the coupling is developed as shown in Figure 4.

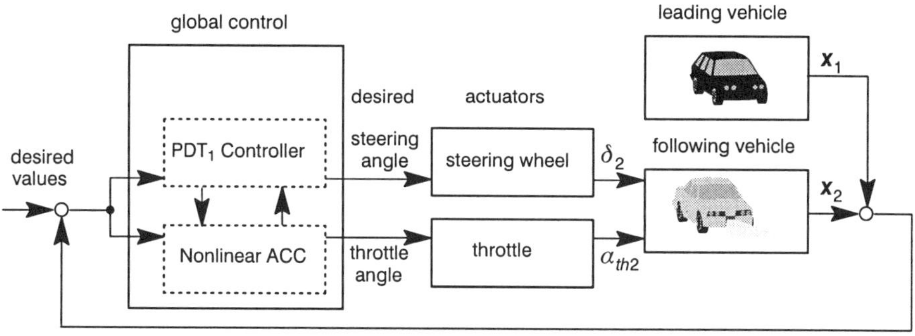

Figure 4: Independent and global control of the vehicle convoy

3.2 Control Strategies

The following measured variables are available: longitudinal and lateral velocity v_{l2} and v_{q2} measured by an optical sensor, longitudinal and lateral acceleration a_{l2} and a_{q2} and the input variables throttle angle α_{th2} and steering angle δ_2 of the controlled vehicle. The relative displacements between the two cars are detected by a laser scanner LMS 200, see [2]. A rotating mirror in the sensor allows 180 deg range measurements of obstacles with an angular resolution of 0.5 deg. The maximum measurement distance is 150 m at a resolution of $0.1m$.

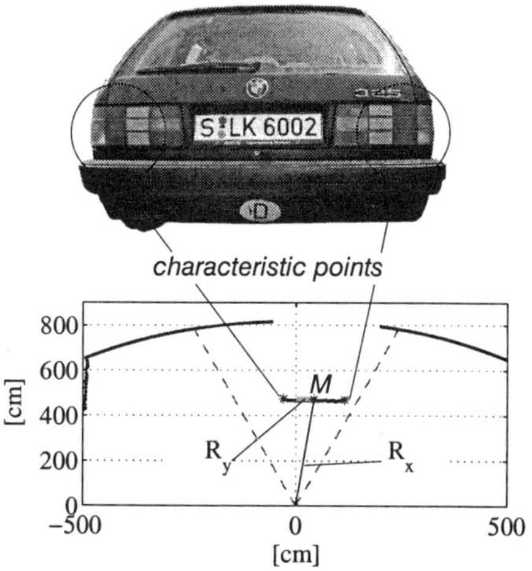

Figure 5: Range measurements

For the range measurements characteristic points are extracted from the measured data, see Figure 5. Then, these points are checked with the length dimensions of the leading vehicle and one gets as a result the distance to the left and right rear edge of the leading car. Using the information of the car edges the relative displacements R_x and R_y to the vehicle in front can be calculated.

Another strategy is storing the coordinates of the point M in a file which results in a discrete path of the leading vehicle, see [6]. In the following, this data is called absolute displacements between the two cars. On the basis of the relative and absolute measurements three different control strategies for the global control of the convoy are proposed.

A. Relative Displacements

The first control strategy uses the relative displacements between the two cars as control variables, see [4]. This approach is shown in Figure 6.

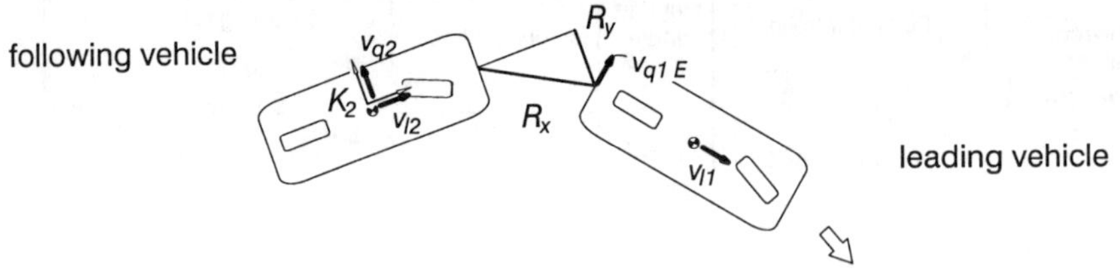

Figure 6: Control variables chosen as relative displacements

The control error of the longitudinal control e_x is the difference between the desired headway R_{xdes} and the relative distance R_x. The desired headway R_{xdes} can be expressed in the form

$$R_{xdes} = d_0 + T_h v_{l2} \ , \tag{4}$$

where d_0 is a constant distance and T_h the headway time. This formulation is called headway control strategy, see [7]. Then, the control error e_{xA} reads as

$$e_{xA} = R_{xdes} - R_x = d_0 + T_h v_{l2} - R_x \ . \tag{5}$$

Differentiating Eq. (5) results in

$$\dot{e}_{xA} = \dot{R}_{xdes} - \dot{R}_x = T_h a_{l2} - (v'_{l1} - v_{l2}) \ , \tag{6}$$

where v'_{l1} is the longitudinal velocity of the leading vehicle transformed into the coordinate system K_2 of the following vehicle.

The lateral control error e_{yA} is the difference between a desired value R_{ydes} which is supposed to be zero and the relative displacement R_y.

$$e_{yA} = R_{ydes} - R_y = -R_y \tag{7}$$

Analogous, one gets the derivative of the lateral control error $\dot{e}_{yA}$ as

$$\dot{e}_{yA} = \dot{R}_{ydes} - \dot{R}_y = -(v'_{q1E} - v_{q2} - \dot{\psi}_2(l_v^* + R_x)) \ , \tag{8}$$

where v'_{q1E} is the lateral velocity of the leading vehicle at the rear bumper transformed into the coordinate system K_2 and l_v^* is the distance between the center of gravity and the front bumper of the following vehicle.

B. Relative and Absolute Displacements

Using the stored trajectory of the leading vehicle the lateral control error can be redefined as the shortest distance R_{traj} between the center of gravity of the following vehicle and the path of the leading vehicle, see Figure 7 and [8].

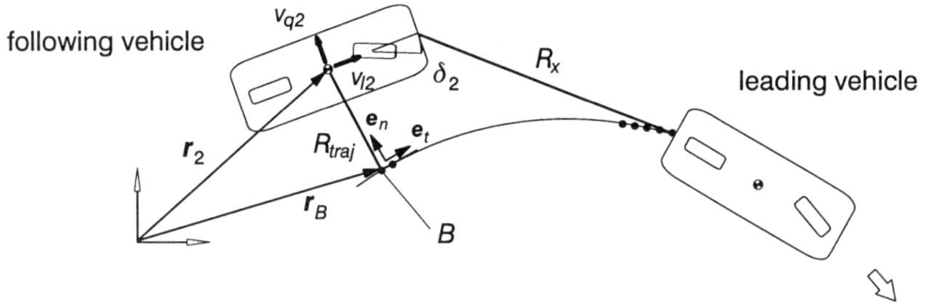

Figure 7: Control variables chosen as relative and absolute displacements

The point B of the path representing the shortest deviation to the following vehicle is determined by a searching algorithm. Then, the tangential and normal vector $\boldsymbol{e}_t$ and $\boldsymbol{e}_n$ of the path in B are calculated using an interpolation. Applying this data, the distance R_{traj} is calculated from the translation vectors $\boldsymbol{r}_2$ and $\boldsymbol{r}_B$ as

$$R_{traj} = [\boldsymbol{r}_2 - \boldsymbol{r}_B]^T \boldsymbol{e}_n \quad . \tag{9}$$

With a desired value $R_{trajdes} = 0$ the lateral control error e_{yB} is given as

$$e_{yB} = R_{trajdes} - R_{traj} = -[\boldsymbol{r}_2 - \boldsymbol{r}_B]^T \boldsymbol{e}_n \quad . \tag{10}$$

The corresponding derivative $\dot{R}_{traj}$ is calculated as the component of the absolute velocity $\boldsymbol{v}_2$ of the following vehicle in normal direction $\boldsymbol{e}_n$,

$$\dot{R}_{traj} = \boldsymbol{v}_2^T \boldsymbol{e}_n = [v_{l2}\ v_{q2}]^T \boldsymbol{e}_n \quad . \tag{11}$$

Then, the derivative of the lateral control error $\dot{e}_{yB}$ reads as

$$\dot{e}_{yB} = \dot{R}_{trajdes} - \dot{R}_{traj} = -[v_{l2}\ v_{q2}]^T \boldsymbol{e}_n \quad . \tag{12}$$

For the longitudinal control, the control variable R_x is used and, therefore, the error e_{xB} and its derivative $\dot{e}_{xB}$ follow from the Eqs. (5) and (6).

C. Absolute Displacements

The third strategy regards only absolute displacements for the calculation of the control errors. While for the the lateral control still the deviation R_{traj} is used, see Eq. (9), for the longitudinal control the range R_x is replaced by the arc length R_{arc} as control variable, see Figure 8.

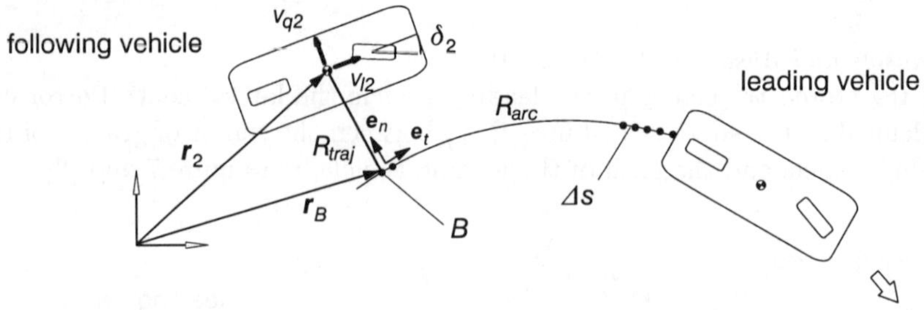

Figure 8: Control variables chosen as absolute displacements

The arc length R_{arc} is composed of linear elements Δs which depend from the scanning time of the laser. Then, the longitudinal error e_{xC} reads as

$$e_{xC} = R_{arcdes} - R_{arc} = d_0 + T_h v_{l2} - \sum \Delta s \ . \tag{13}$$

The corresponding derivative of the arc length $\dot{R}_{arc}$ is calculated from the difference between the longitudinal velocity of the leading vehicle v_{l1} and the component of the absolute velocity $\boldsymbol{v}_2$ of the following vehicle in tangential direction $\boldsymbol{e}_t$,

$$\dot{R}_{arc} = v_{l1} - \boldsymbol{v}_2^T \boldsymbol{e}_t = v_{l1} - [v_{l2} \ v_{q2}]^T \boldsymbol{e}_t \ . \tag{14}$$

Then, the derivative of the longitudinal control error $\dot{e}_{xC}$ is given as

$$\dot{e}_{xC} = \dot{R}_{arcdes} - \dot{R}_{arc} = T_h a_{l2} - v_{l1} + [v_{l2} \ v_{q2}]^T \boldsymbol{e}_t \ . \tag{15}$$

On the basis of the defined control errors $e_{x/yj}$, $j = A, B, C$, a nonlinear control design is presented in the next section.

3.3 Control Design

The control design is divided into three steps. The first one is a state transformation called input-output-linearization, see [9]. The second step is the stabilization of the control errors and finally the so-called internal dynamics has to be examined guaranteeing the stability of the global controller.

Input-Output-Linearization
The first step of the control design is the compensation of the nonlinearities and the decoupling of the longitudinal and lateral motion using a state space transformation $\boldsymbol{z} = [\boldsymbol{z}^{e^T} \boldsymbol{z}^{i^T}]^T$. With the outputs chosen as

$$y_1 = x_{l2} \text{ and } y_2 = x_{q2} \ . \tag{16}$$

the transformation $\boldsymbol{z}^e$ is given by x_{l2} and x_{q2} and its derivatives in the inertial system v_{l2}, a_{l2} and v_{q2}, which leads to

$$\dot{\boldsymbol{z}}^e = \begin{bmatrix} \dot{z}^e_{11} \\ \dot{z}^e_{12} \\ \dot{z}^e_{13} \\ \dot{z}^e_{21} \\ \dot{z}^e_{22} \end{bmatrix} = \begin{bmatrix} z^e_{12} \\ z^e_{13} \\ \pi_1(\boldsymbol{z}^e, \alpha_{th2}) \\ z^e_{22} \\ \pi_2(\boldsymbol{z}^e, \boldsymbol{z}^i, \delta_2) \end{bmatrix} . \tag{17}$$

The functions $\pi_1(\boldsymbol{z}^e, \alpha_{th2})$ and $\pi_2(\boldsymbol{z}^e, \boldsymbol{z}^i, \delta_2)$ are given by

$$\begin{aligned} \pi_1(\boldsymbol{z}^e, \alpha_{th2}) &= \frac{k_m}{T_1 r m_l}\left(m_0 + m_1 z^e_{12} + m_2 \alpha_{th2}\right) - 2\frac{k_l}{m_l} z^e_{12} z^e_{13} \\ &- \frac{1}{T_1}\left(m_l z^e_{13} + k_r + k_c + k_l z^{e2}_{12}\right) \qquad (18) \\ \pi_2(\boldsymbol{z}^e, \boldsymbol{z}^i, \delta_2) &= \frac{1}{m_q}\left(\frac{z^i_{32}}{z^e_{12}}(C_h l_h - C_v l_v) - \frac{z^e_{22}}{z^e_{12}}(C_h + C_v) + \delta_2(C_v - k_{rv})\right), \qquad (19) \end{aligned}$$

where the mass m_l is an abbreviation for $m_q + 2(j_r + j_{red})/r^2$. The second part of the transformation, the so-called internal dynamics $\boldsymbol{z}^i = [\psi_2 \ \dot{\psi}_2]^T$, is given as

$$\dot{\boldsymbol{z}}^i = \begin{bmatrix} \dot{z}^i_{31} \\ \dot{z}^i_{32} \end{bmatrix} = \begin{bmatrix} z^i_{32} \\ \pi_3(\boldsymbol{z}^e, \boldsymbol{z}^i, \delta_2) \end{bmatrix} \tag{20}$$

and describes the equations of motion of the yaw angle ψ_2. To compensate the nonlinearities given by the Eqs. (18) and (19) two new inputs

$$w_1 = \pi_1(\boldsymbol{z}^e, \alpha_{th2}) \quad \text{and} \quad w_2 = \pi_2(\boldsymbol{z}^e, \boldsymbol{z}^i, \delta_2) \tag{21}$$

are introduced. Now, it is possible to calculate a linearized control law by solving Eqs. (21) for $\alpha_{th2} = \pi_1^\star(\boldsymbol{z}^e, w_1)$ and $\delta_2 = \pi_2^\star(\boldsymbol{z}^e, \boldsymbol{z}^i, w_2)$:

$$\begin{aligned} \alpha_{th2} &= \frac{1}{m_2}\left(\frac{rT_1}{k_m}(w_1 m_l + 2k_l z^e_{12} z^e_{13}) - m_0 - m_1 z^e_{12}\right) \\ &+ \frac{r}{k_m m_2}(m_l z^e_{13} + k_r + k_c + k_l z^{e2}_{12}) \ , \qquad (22) \\ \delta_2 &= \frac{1}{C_v - k_{rv}}\left(m_q w_2 + \frac{z^e_{22}}{z^e_{12}}(C_h + C_v) - \frac{z^i_{32}}{z^e_{12}}(C_h l_h - C_v l_v)\right) , \qquad (23) \end{aligned}$$

which yields as a result of the compensation two decoupled linear systems in Brunovsky canonical form, see [9],

$$y_1^{(3)} = w_1 \quad \text{and} \quad \ddot{y}_2 = w_2 \tag{24}$$

and the internal dynamics

$$\dot{z}^i = \begin{bmatrix} z^i_{32} \\ \pi_3\left(z^e, z^i, \pi_2^\star(z^e, z^i, w_2)\right) \end{bmatrix} . \tag{25}$$

The system representation of the control design model after the input-output-linearization is shown in Figure 9 consisting of the Brunovsky canonical form and the internal dynamics.

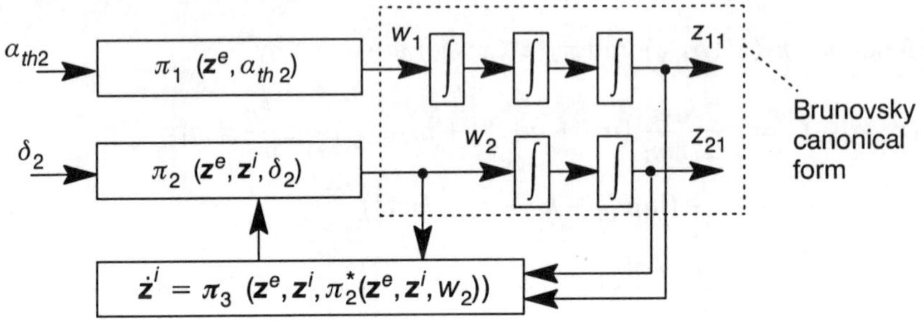

Figure 9: System representation after the input-output-linearization

The result of the state space transformation z is the compensation of the nonlinearities and the decoupling of the lateral and the longitudinal motion. The next step of the control design is the stabilization of the control errors in both directions which will be achieved by a tracking formulation.

Tracking Control
Regarding tracking control the errors $e_{x/yj}$, $j = A, B, C$, defined in section 3.2 are to be minimized by changing the outputs along desired trajectories. The differential equations of the tracking errors in both direction are given by

$$\ddot{e}_{x/yj} + \sum_{k=0}^{1} \ell_{x/yk}\, e^{(k)}_{x/yj} = 0 \qquad j = A, B, C \ . \tag{26}$$

Properly chosen coefficients $\ell_{x/yk}, k = 0, 1$ provide stabilization of the control errors. By solving the Eqs. (26) for $y_1^{(3)}$ and $\ddot{y}_2$, it follows from Eq. (24)

$$w_1 = \frac{1}{T_h}\left[R_j^{(3)} - \sum_{k=0}^{1} \ell_{xk}\, e^{(k)}_{xj}\right] \text{ and } w_2 = -\sum_{k=0}^{1} \ell_{yk}\, e^{(k)}_{yj} \quad j = A, B, C \ . \tag{27}$$

With the Eqs. (27) the feedback laws in Eq. (22) and (23) result in

$$\alpha_{th2} = \pi_1^\star\left(z^e, \frac{1}{T_h}\left[R_j^{(3)} - \sum_{k=0}^{1} \ell_{xk}\, e^{(k)}_{xj}\right]\right) \text{ and } \delta_2 = \pi_2^\star\left(z^e, z^i, -\sum_{k=0}^{1} \ell_{yk}\, e^{(k)}_{yj}\right) , \tag{28}$$

consisting of the compensation given by the transformation z and the error stabilization.

Zero Dynamics
The zero dynamics describes the remaining motions of the controlled system if all chosen outputs, Eq. (16), their derivatives and the new introduced inputs, Eq. (21), are zero, see [9]. Since there is no control force any more, this motion has to be stable. One gets the zero dynamics from the internal dynamics, Eq. (25), by replacing the constraints $z^e = w_2 = 0$ as

$$\left(j_z + m_r(l_v^2 + l_h^2)\right)\ddot{\psi}_2 + \frac{\dot{\psi}_2}{v_{l2}}\left(l_h^2 C_h + l_h l_v C_h\right) = 0 \ . \quad (29)$$

Regarding the eigenvalue criterion, the motion of the yaw rate $\dot{\psi}_2$ is stable, since $l_h C_h (l_h + l_v)$ is always greater zero. Therefore, the zero dynamics is always stable as far as the control design model, Eq. (1), is valid.

In the next section the global control design is compared with the two independent controller by simulation.

4 SIMULATION RESULTS

The maneuver of the leading vehicle for the following simulations is an accelerated double lane change. The longitudinal velocity v_{l1} and the trajectory of the leading car are shown in Figure 10. Because of the step disturbance in the

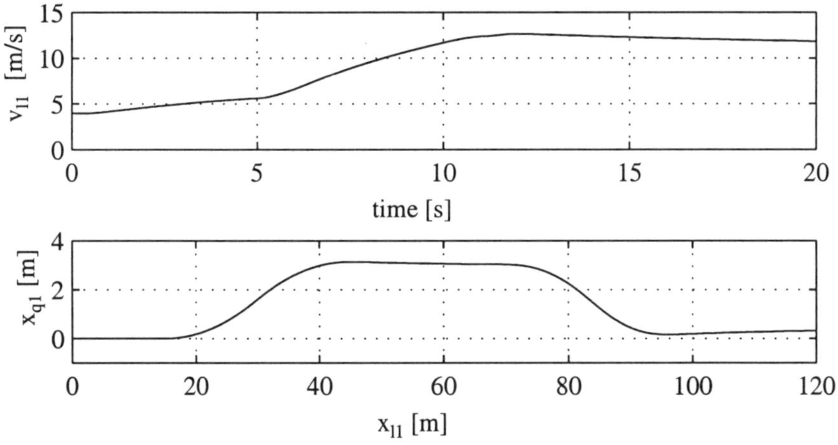

Figure 10: Longitudinal velocity and trajectory of the leading vehicle

throttle position between $\alpha_{th1} = 5$ deg to $\alpha_{th1} = 18$ deg the leading vehicle accelerates from $4m/s$ to $13m/s$. The following vehicle travels for $t = 0s$ with a constant velocity of $3m/s$. There is no initial headway error, the desired range is calculated with the constant distance $d_0 = 5m$ and the headway time $T_h = 1s$.

In the following, the independent and the global controller are compared using the control strategy A. The control parameter of the global controller are $l_{x0} = 20$, $l_{x1} = 14$, $l_{y0} = 25$ and $l_{y1} = 10$, while the lateral PDT_1 controller is given by $P = P_4 = 17$, $T_d = 0.1$ and $T_1 = 0.005$, see [4]. For the ACC the same parameter as for the global controller in longitudinal direction are chosen.

4.1 Strategy A - Relative Displacements

Before the simulation results are compared, the lateral control stability is evaluated on the basis of the Bode plot in Figure 11.

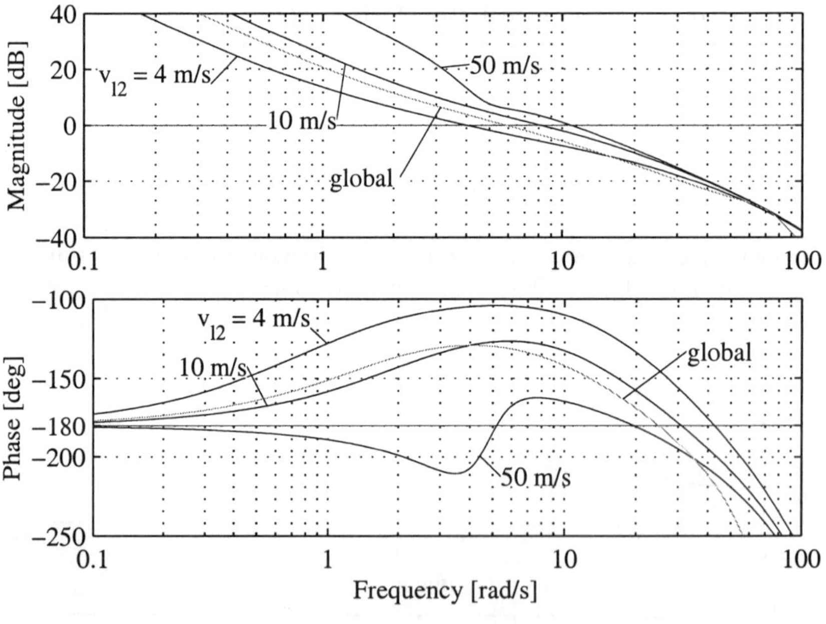

Figure 11: Lateral control stability of the PDT_1 and the global controller

One recognizes that the phase margin γ of the PDT_1 controller is reduced with increasing longitudinal velocity v_{l2} while the phase margin of the global controller is independent from the longitudinal velocity.

The simulated lateral control error e_{yA}, pictured in Figure 12, clarifies that there are only small differences between the global and the independent controller with the gain P_4. Since the lateral control stability of the independent controller depends from the longitudinal velocity, see Figure 11, a second result of the PDT_1 controller is plotted in Figure 12, where the controller gain P_{13} is chosen in such a way that the phase margins γ_{P_4} and $\gamma_{P_{13}}$ for the two different velocities $v_{l2} = 4m/s$ and $v_{l2} = 13m/s$ resemble $\gamma_{P_4} = \gamma_{P_{13}}$. Then, it turns out that the lateral error e_{yA} of the independent controller with the new gain P_{13} becomes larger. Therefore, the global control concept has the structural advantage that the phase margin is constant and therewith a velocity depending control parameter adjustment can be avoided. For this reason, only the global controller is considered for the comparison of the other control strategies.

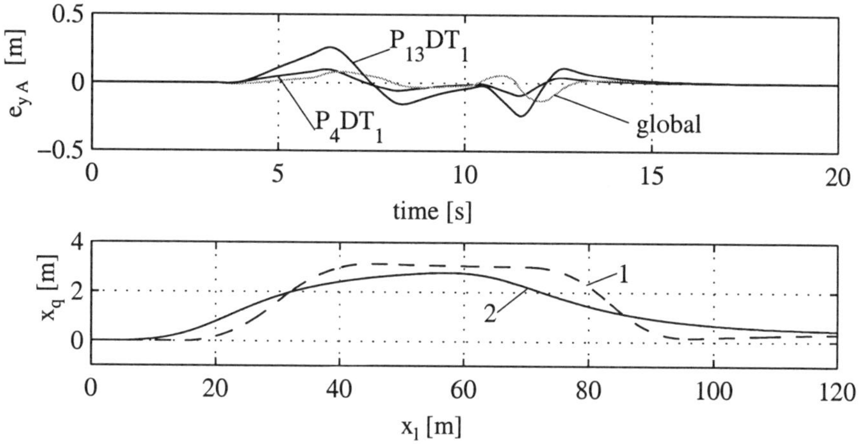

Figure 12: Strategy A - lateral control error and trajectories of both vehicles

Comparing the trajectories of the two vehicles, see Figure 12, one notices that although the lateral control error is very small, the following vehicle cuts the path of the leading one. Therefore, the control strategy A using relative displacements as control variables is not suited for trajectory following.

4.2 Strategy B - Relative and Absolute Displacements

The Figure 13 shows the simulation results of the trajectories and the lateral control error using the control strategy B. It turns out that the trajectories of the two vehicle agree sufficiently well where the lateral control error is also small. On the other hand, regarding the throttle angle and the longitudinal control error in Figure 14, one recognizes an oscillation in the throttle as well as in the control error what is also not acceptable.

The reason for the oscillation is the choice of the control variables. In the case of vanishing lateral control error, the controlled vehicle follows the leading one within the range R_x representing the secant to the leading vehicle. On the other hand, the vehicle has to move the longer way on the arc length as demand of the lateral control resulting in an oscillation of the throttle.

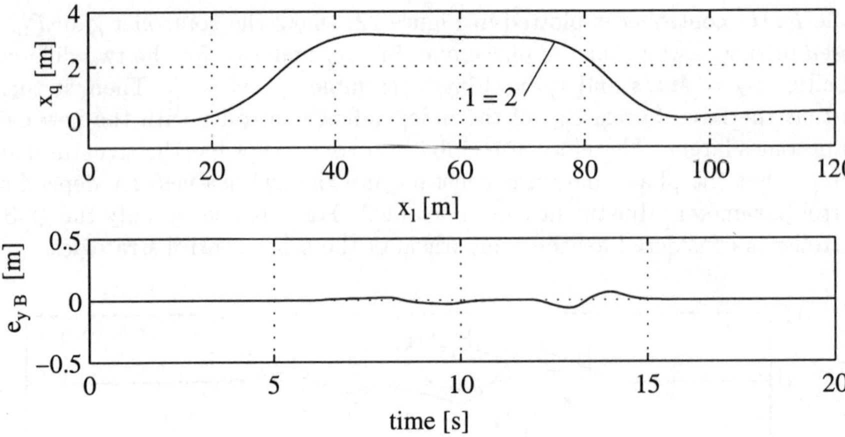

Figure 13: Strategy B - trajectories of both vehicles and lateral control error

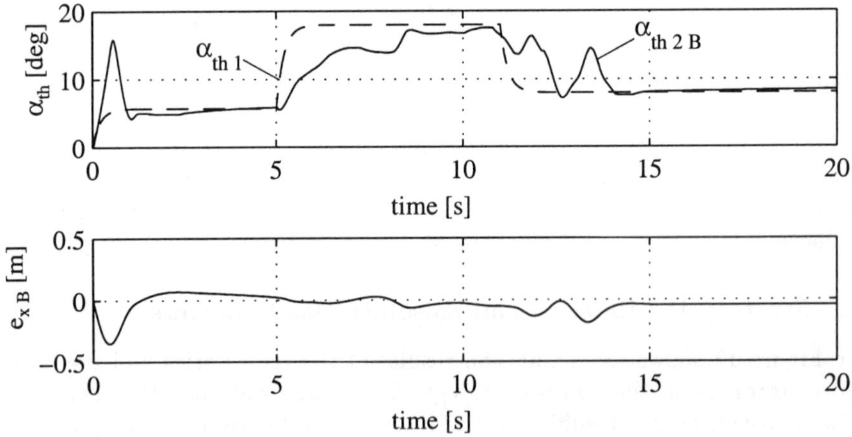

Figure 14: Strategy B - throttle angle of both cars and longitudinal control error

4.3 Strategy C - Absolute Displacements

For a comparison, now the control strategies B and C are shown in one plot. Figure 15 displays the trajectories of the two vehicles and the lateral control error. As expected, nothing changes in the behavior of the lateral control. The results are nearly identical.

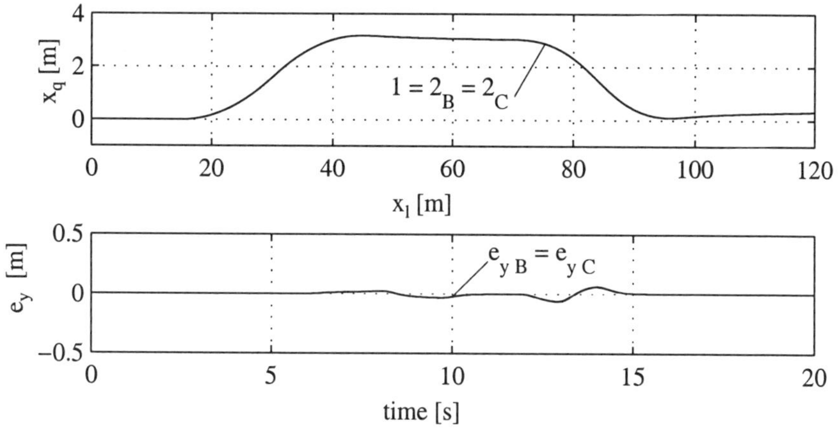

Figure 15: Strategy B and C - trajectories and lateral control error

The simulation results of the throttle angle and the longitudinal control error are shown in Figure 16. One recognizes that the oscillation in the throttle is reduced as well as in the control error applying the strategy C.

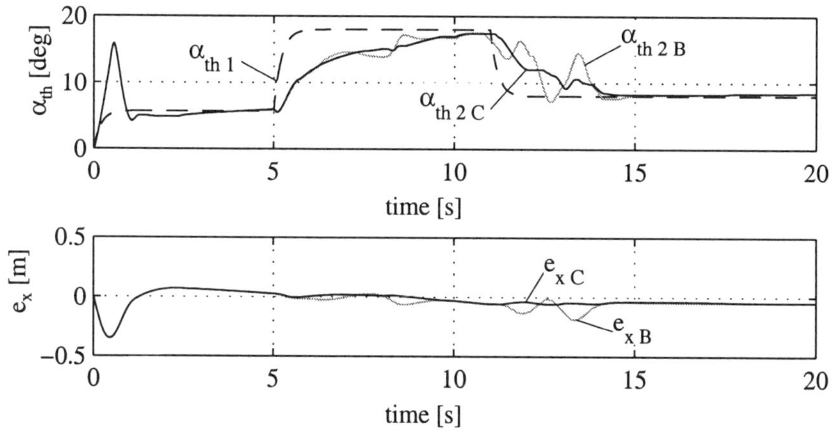

Figure 16: Strategy B and C - throttle angle and longitudinal control error

5 CONCLUSION

In this paper, a global tracking control is developed for the lateral and longitudinal dynamics of a vehicle convoy. This controller is compared with two independently designed controllers from a previous work - a linear PDT_1 controller for the lateral direction and an ACC for the longitudinal motion - by simulation. It turns out that the global control avoids the velocity depending control parameter adjustment which is necessary for the independently designed controller.

To achieve the control objective of path following different control strategies with relative and absolute control variables are compared. It is shown that path following using only relative vehicle displacements is not feasible. Using absolute displacements only in lateral direction results in throttle oscillation and considerable longitudinal control error due to the cornering. In contrast to that, applying absolute displacements in both directions good results are obtained.

REFERENCES

1. Petersen, U.; Schiehlen, W., "Lateral Control of a Convoy Vehicle system", Proc. 14th IAVSD, Ann Arbor, USA, 21–25 August 1995, pp. 519–532.
2. Fritz, A.; Schiehlen, W., "Nonlinear ACC in Simulation and Measurement", Proc. 5th AVEC, Ann Arbor, USA, 22–24 August 2000, pp. 15–22.
3. Schiehlen, W., "Multibody System Dynamics: Roots and Perspectives", Multibody System Dynamics, Vol.1, No.2, 1997, pp. 149–188.
4. Petersen, U., "Zur Dynamik eines mechatronisch gekoppelten Fahrzeuggespanns", VDI Verlag, 1998.
5. Popp, K.; Schiehlen, W., "Fahrzeugdynamik", Teubner Verlag, 1983.
6. Gehrig, S.; Stein, F., "A Trajectory-Based Approach for the Lateral control of Vehicle Following Systems", Proc. IV 98, Stuttgart, Germany, 1998, pp. 156–161.
7. Swaroop, D. et al, "A Comparison of Spacing and Headway Control Laws for Automatically Controlled Vehicles", Vehicle System Dynamics, Vol. 23, 1994, pp. 597-625.
8. Pham, H.; Hedrick, K.; Tomizuka, M., "Autonomous Steering and Cruise Control of Automobiles via Sliding Mode Control", Proc. 3rd AVEC, Tsukuba, Japan, 1994, pp. 444–448.
9. Isidori, A., "Nonlinear Control Systems", Springer-Verlag, 1989.

Vehicle System Dynamics Supplement 35 (2001), pp. 165-180

Some Advanced Control Techniques for Automatic Steering of Road Vehicles

RICARDO A. RAMIREZ M.*, MOHAMMED M'SAAD† and LUC DUGARD‡

SUMMARY

This paper presents a feasibility study of the robust predictive control with and without parameter adaptation for the design of an automatic steering system. The involved vehicle is a full-scale nonlinear model and the feedback is only on the lateral displacement unlike in the available studies where an additional feedback on the yaw rate is commonly used to improve the performance. The robustness feature is particularly motivated by variations in velocity, tyre-road contact and vehicle mass. The underlying control design is performed using a sensitivity function shaping oriented procedure that heavily borrows from the robust control culture. Intensive simulations are carried out to emphasize the performance of the proposed automatic steering system.

1 INTRODUCTION

Advanced control of automotive vehicle systems is more and more discussed in the open literature and widely developed in industry. It has been pointed out that automatic steering is of fundamental interest for urban transport vehicles and Automated Highway Systems (AHS) of the next century, see, e.g., [1]. One of the fundamental parts of AHS are lane following manœuvres, e.g., automatic steering. The main purpose of automatic steering consists in performing a robust tracking of a road curvature in spite of uncertain operating conditions due to large variations in velocity, contact characteristics between tyres and road surface and vehicle mass. The underlying control systems usually involve feedbacks of both the lateral displacement and the yaw rate, and use the front steering angle as a manipulated variable. The reference is located in the middle of the lane and may consist of permanent magnets in the road or the magnetic

*Department of Mechatronics and Automation, DIA, ITESM Campus Monterrey, E. Garza Sada 2501 Sur, Monterrey N.L. C.P. 64849, México

†Laboratoire d'Automatique et de Procédés, ISMRA, 6 Blvd. Maréchal Juin, 14050 Caen Cedex, France

‡Laboratoire d'Automatique de Grenoble, (UMR 5528 CNRS-INPG-UJF), ENSIEG, BP 46, 38402 St Martin d'Hères, France

field of an electrically supplied wire or a discrete marker scheme. Alternatively a vision-based lane sensing scheme could be used, see, e.g., [2]. The involved tracking error is measured by a displacement sensor mounted in the center of the front end of the vehicle which can be obtained from measurements of the relative strength of magnetic fields emitted by permanent magnets or discrete magnetic markers. The yaw rate measurement is performed by a yaw rate sensor.

Automatic steering has been extensively studied during the last ten years. For instance, it has been showed that the decoupling of lateral and yaw motion by yaw-rate feedback to front-wheel steering, leading to a significant reduction of the lateral deviation compared to earlier controller design which used solely feedback of the lateral deviation, see, e.g., [3]. In [4] designed a nonlinear sliding mode controller that showed good lane keeping ability and robustness to parameter variations while maintaining passenger comfort. Here, the control objective is to achieve a fixed dynamic relationship between the driver steering commands and the vehicle yaw rate in spite of variations in the operating conditions. The automatic steering control problem represents a challenging opportunity to investigate advanced control techniques that have reached a reasonable level of maturity. This is mainly a result of many years of effort devoted to the understanding of the control theory as well as the tremendous progress in the computer technology that makes the implementation of control systems easier and cheaper.

In the present paper, the authors aim at investigating the applicability of the robust predictive control with and without suitable parameter adaptation for the design of an automatic steering system using only lateral displacement measurement unlike in earlier design studies, see, e.g., [5]. A realistic simulation framework involving a full-scale nonlinear model of a BMW520i vehicle is used to this end. This simulator has been developed at the Institute for Industrial Information Techniques of the Technical University Karlsruhe, to investigate automotive control problems, see, e.g., [6]. More specifically, the underlying control design is carried out using an appropriate predictive control approach, developed from the generalized predictive control in the spirit of the robust control theory, see, e.g., [7] and [8].

Besides the standard design specifications, namely maintaining the lateral displacement as small as possible for typical manœuvres, taking into account the actuator constraints and the passenger comfort, suitable shapes of the usual sensitivity functions have to be achieved. This allows to deal with those ubiquitous variations in velocity, road adhesion factor and vehicle mass. The usual sensitivity functions are used as suitable quantifiers for both nominal performance and stability robustness. It is worth mentioning that the adopted methodology requires a good know-how in system identification as the control system and its usual sensitivity functions are determined from control models that should be identified over the domain of possible operating conditions.

The involved experimental evaluation is carried out using the advanced control software package SIMART which allows to perform all the steps involved in any genuine control system design. This more detailed perspective is outlined in,

e.g., [9].
The paper is organized as follows: the full-scale nonlinear model for vehicle dynamics is presented in Section 2. In Section 3 the predictive control approach is developed as well as its adaptive version. A realistic simulation study is performed in Section 4. Some concluding remarks are mentioned in the last Section.

2 FULL-SCALE NONLINEAR MODEL

In the present feasibility study, a full-scale nonlinear vehicle model, which has been developed at the Institute for Industrial Information Techniques of Karlsruhe, is used, see, e.g., [10]. This consists of several submodels e.g. chassis, traction/braking, suspension, road model, vehicle aerodynamic and tyres and takes all important nonlinearities into account. The global inputs for the involved simulator are the front steering angle, the traction or braking torques for each wheel, the initial velocity of the vehicle, the weather and road conditions and the road path. The latter is composed of different road section for each them the desired velocity needs to be defined, as well as the road curvature, the wind speed and direction, the weather conditions, e.g. dry, wet, snow surface. Figure 1 shows the submodels which have been implemented using the software Matlab environment. The data have been adapted to a BMW520i and proved to demonstrate a realistic behaviour during driving conditions, see, e.g., [11].

The chassis model is assumed to be represented by a three dimensional body with external forces acting on it because of the tyre-road contact, the aerodynamic effects and the gravity. The expressions governing the translational and rotational motions of the chassis could be derived from Newton's third law of motion and Euler's equations respectively applied in the direction of three axis. A complex nonlinear tyre model is used to generate the longitudinal and lateral forces, see, e.g., [12], [13] and [14]. A number of factors affect the tyre dynamic: the normal forces acting on the tyre, the tyre-road characteristic and the forces generated by braking and acceleration actions. In this study, the longitudinal and lateral forces acting on the front tyres are described by the nonlinear model. Previous experimental studies have demonstrated the effectiveness of the proposed tyre model to describe the dynamic behaviour of the longitudinal and lateral tyre forces.

3 PREDICTIVE CONTROL APPROACH

In this section we develop the predictive control approach which yields the control algorithm used for automatic steering. The predictive control design is based on the assumption that input-output behaviour of the plant to be

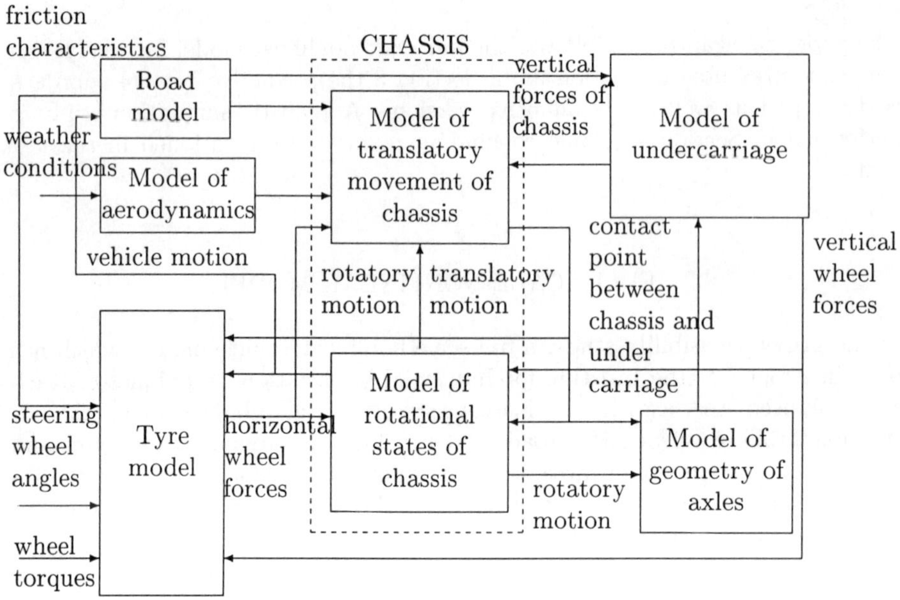

Figure 1: Submodels of the vehicle

controlled can be appropriately approximated by the following backward shift operator (q^{-1}) model

$$\begin{aligned} A(q^{-1})y(t) &= B(q^{-1})u(t-d-1)+v(t) \\ D(q^{-1})v(t) &= C(q^{-1})\gamma(t) \end{aligned}$$

with

$$\begin{aligned} A(q^{-1}) &= 1+a_1q^{-1}+...+a_{na}q^{-na} \\ B(q^{-1}) &= b_o+b_1q^{-1}+...+b_{nb}q^{-nb} \\ C(q^{-1}) &= 1+c_1q^{-1}+...+c_{nc}q^{-nc} \\ D(q^{-1}) &= 1+d_1q^{-1}+...+d_{nd}q^{-nd} \end{aligned}$$

where $u(t)$ is the control variable, $y(t)$ is the measured plant output, d denotes the minimum plant model delay in sampling periods, $v(t)$ represents the external disturbances and $\{\gamma(t)\}$ is assumed to be a sequence of widely spread pulses of unknown magnitude or independent random variables with zero mean values and finite variance. The predictive control objective consists in minimizing, in a receding horizon sense with respect to the vector input $U_f(t+ch-1) = [u_f(t)\dots u_f(t+ch-1)]^T$, the following linear quadratic cost function

$$E\left\{\sum_{j=sh}^{ph}(y_f(t+j))^2+\rho(u_f(t+j-sh))^2\right\}$$

subject to the constraints

$$u_f(t+i) = 0 \quad \text{for} \quad ch \le i < ph$$

with

$$\begin{aligned} W_{yd}(q^{-1})y_f(t) &= W_{yn}(q^{-1})y(t) \\ W_{ud}(q^{-1})u_f(t) &= D(q^{-1})W_{un}(q^{-1})u(t) \end{aligned}$$

where $E\{\cdot\}$ denotes the mathematical expectation. *sh*, *ph* and *ch* are the starting, prediction and control horizons according to the long range predictive control culture, ρ is a positive scalar. $\mathcal{W}_u(z^{-1}) = \frac{D(z^{-1})W_{un}(z^{-1})}{W_{ud}(z^{-1})}$ and $\mathcal{W}_y(z^{-1}) = \frac{W_{yn}(z^{-1})}{W_{yd}(z^{-1})}$ denote user specified input and output frequency weightings, respectively. The frequency weightings are mainly motivated by stability and performance robustness considerations and are such that all polynomials $W_{xx}(q^{-1})$ are Hurwitz. The underlying control problem will be handled using the generalized predictive control approach proposed in [8] from the following plant reparametrization:

$$\bar{A}(q^{-1})y_f(t) = \bar{B}(q^{-1})u_f(t-d-1) + \bar{C}(q^{-1})\gamma(t)$$

with

$$\begin{aligned} \bar{A}(q^{-1}) &= A(q^{-1})D(q^{-1})W_{yd}(q^{-1})W_{un}(q^{-1}) \\ \bar{B}(q^{-1}) &= B(q^{-1})W_{yn}(q^{-1})W_{ud}(q^{-1}) \\ \bar{C}(q^{-1}) &= A(q^{-1})C(q^{-1})W_{yn}(q^{-1})W_{un}(q^{-1}) \end{aligned}$$

The resulting controller may be given the following linear form:

$$S(q^{-1})D(q^{-1})u(t) + R(q^{-1})y(t) = 0$$

with

$$\begin{aligned} S(q^{-1}) &= \bar{S}(q^{-1})W_{yd}(q^{-1})W_{un}(q^{-1}) \\ R(q^{-1}) &= \bar{R}(q^{-1})W_{yn}(q^{-1})W_{ud}(q^{-1}) \end{aligned}$$

where the polynomials $\bar{S}(q^{-1})$ and $\bar{R}(q^{-1})$ depend on the plant model as well as on the design parameters. The involved nominal control system may be represented as shown in figure 2 where $v_u(t)$ and $v_y(t)$ respectively denote the input and output disturbances, and $\nu_u(t)$ and $\nu_y(t)$ the input and output noise measurements. The corresponding input-output behaviour is described by

$$\begin{aligned} P_c(q^{-1})y(t) &= q^{-d-1}B(q^{-1})S(q^{-1})D(q^{-1})(v_u(t) + \nu_u(t)) \\ &+ A(q^{-1})S(q^{-1})D(q^{-1})v_y(t) - q^{-d-1}B(q^{-1})R(q^{-1})\nu_y(t) \end{aligned}$$

$$\begin{aligned} P_c(q^{-1})u(t) &= A(q^{-1})S(q^{-1})D(q^{-1})v_u(t) - q^{-d-1}B(q^{-1})R(q^{-1})\nu_u(t) \\ &- A(q^{-1})R(q^{-1})[v_y(t) + \nu_y(t)] \end{aligned}$$

The control system is asymptotically stable if and only if $P_c(q^{-1})$ is Hurwitz.

$$P_c(q^{-1}) \;=\; 0 \;\Rightarrow\; |q| < 1$$

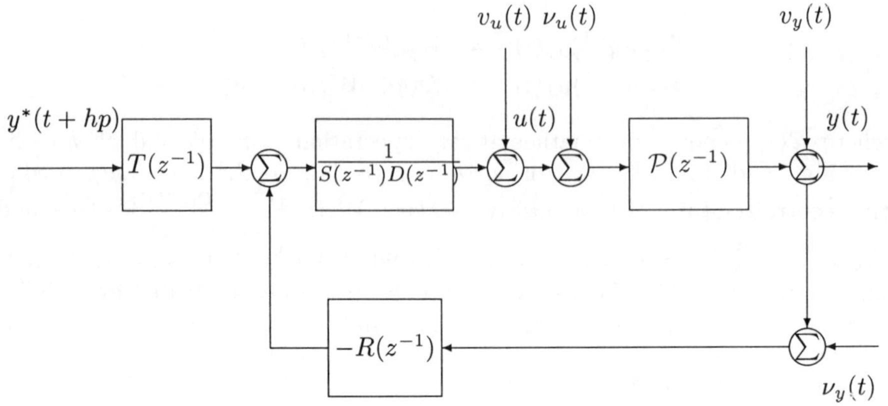

Figure 2: Nominal control system

3.1 The Adaptive Controller

The key issues to get a robust adaptive controller are suitable signal processing and a robust parameter adaptation algorithm, see, e.g., [15]. The former can be simply performed from the plant model by filtering and normalizing the data, while the latter has to incorporate an adequate parameter adaptation alertness and freezing. These robustness features can be achieved by the regularized constant trace algorithm with conditional parameter adaptation. More specifically, the parameter adaptation is frozen whenever the available information is not likely to improve the parameter estimation process. The adaptive control law is obtained by simply invoking the certainty equivalence principle. The admissible estimated model is obtained from the involved parameter adaptation algorithm. The adaptive control system is implemented as follows:

1. Wait for the clock pulse and sample the plant output.

2. Update the plant model parameters using the considered parameter estimator.

3. Construct the admissible estimated plant model as follows:

$$\theta_a(t) = \begin{cases} \hat{\theta}(t) & \text{if } \hat{\theta}(t) \text{ is stabilizable} \\ \theta_a(t-1) & \text{otherwise} \end{cases}$$

   where $\hat{\theta}(t)$ is the estimated plant model provided by the considered parameter adaptation algorithm.

4. Evaluate the adaptive control law using the admissible estimated plant model $\theta_a(t)$.

5. Implement the control signal and go to 1.

The nominal performance of the control system as well as the stability robustness can be evaluated from its usual sensitivity functions relating the exogenous inputs to the plant input and output, respectively. The shapes of the sensitivity functions may be refined by properly specifying the involved design parameters, namely the frequency weightings, the input weighting ρ, the observer dynamics P_0, the starting sh, prediction ph and control ch horizons. To do so, the advanced control software package SIMART has been used.

4 EXPERIMENTAL EVALUATION

This section is devoted to the investigation of the applicability of the proposed predictive control approach in a realistic simulation framework involving a full-scale nonlinear model of a BMW520i. The main purpose of automatic steering is to track a given road curvature path. The error between the guideline and the vehicle $y_s(t)$ is assumed to be measured by a displacement sensor mounted in the front end of the vehicle which can be obtained from measurements of the relative strength of magnetic fields by permanent magnets or discrete magnetic markers, the control variable is the front steering angle $\delta_f(t)$. The control objective consists in maintaining the lateral displacement as small as possible in spite of variations in road curvature. Of particular importance, the controller should perform well in the presence of variations in mass, velocity and contact between tyre and road surface. Figure 3 shows the vehicle and the variables of interest.

4.1 Design Specifications

The design specifications have been adapted from those considered in earlier design studies, They are primarily given in terms of actuator constraints and passenger comfort as follows:

- The steering angle and its rate should satisfy $|\delta_f(t)| \leq 40$ deg and $|\dot{\delta}_f(t)| \leq 45$ deg/s.

- The displacement from the guideline must not exceed 15 cm in transient behaviour and 2 cm in steady state behaviour.

- The lateral acceleration should satisfy

$$|a_y(t)| \leq \frac{v(t)^2}{\rho(t)} \pm 0.1g$$

  where $v(t)$, $\rho(t)$ and g denote the velocity, the curvature radius and the gravity constant, respectively.

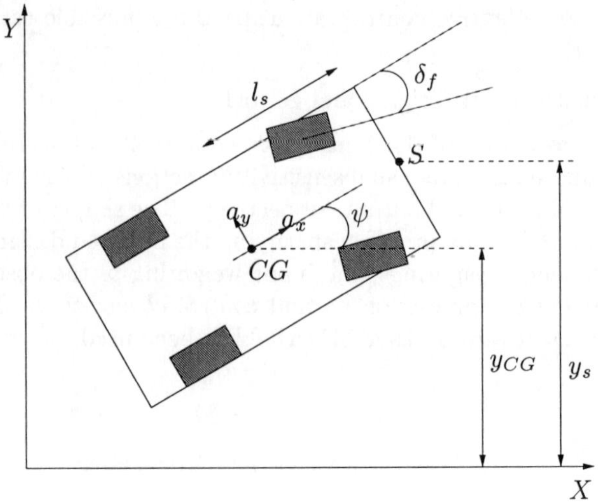

Figure 3: Vehicle and pertinent variables

4.2 Reference Driving Manœuvres

Three reference driving manœuvres are considered to evaluate the performance of the automatic steering system. The first consists in a transition from a straight line into a circular one with a radius of 400 m, the second involves a wind gust disturbance at 20 m/s in lateral direction. The third simulation consists in a transition from manual mode to automatic steering. For the sake of space limitations, only the first driving manœuvre is reported.

4.3 Off-line Identification Procedure

As the predictive control approach is based on linear control design model, the vehicle dynamics have been identified over an important domain of driving conditions, which is shown in figure 4. The identification has been performed using

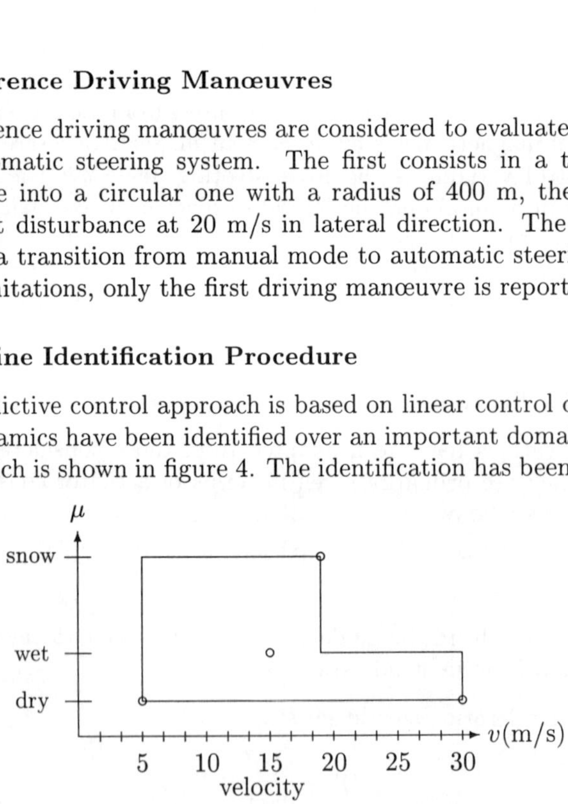

Figure 4: Operating domain

a least squares method incorporating suitable signal filtering, data normalization, adaptation gain regularization and any prior knowledge. The data are the front steering angle $\delta_f(t)$ and the second derivative of the sensor displacement $\ddot{y}_s(t)$. The latter has been determined using the lateral acceleration of the vehicle $a_y(t)$ and the second derivative of the yaw angle $\ddot{\psi}(t)$, as follows

$$\ddot{y}_s(t) \quad = \quad (a_y(t) + l_s\ddot{\psi}(t)) \cos \psi(t) \simeq a_y(t) + l_s\ddot{\psi}(t)$$

where l_s denotes the distance between the centre of gravity and the sensor position. This formula is valid for a straight path which has been considered for all identification experiments. The lateral acceleration can be directly measured by an accelerometer while the second derivative of the yaw angle can be obtained either by derivation of the measurable yaw rate or by an additional accelerometer mounted at the front-end of the vehicle. In this case, the following equation can be used:

$$\ddot{\psi}(t) \quad = \quad \frac{a_{ys}(t) - a_y(t)}{l_s}$$

The control design models have been obtained from the identified model by simply multiplying by a double integrator. This provides sets of pulse transfer functions relating the sensor displacement to the steering angle.

4.4 Control Model Selection

Intensive identification experiments, using suitable offset-free binary random sequences, have been made to determine a nominal model for robust control design. The identified models show that the crucial variations are those of the velocity and road adhesion factor. The dynamics of the vehicle are not very sensitive to variations in mass. An average mass corresponding to the mass of the vehicle loaded with the driver and one passenger has been considered throughout the experimental evaluation. The Bode plots of the identified models obtained for various velocities on a dry road are shown in figure 5. Figure 6 shows the Bode plots of the models on a dry, a wet and on snow road, respectively, for a speed of 15 m/s. The identified models have been validated both in the time and frequency domains. In the time domain, the identification performance quantifiers are based on correlation function, information likelihood measure and singular values of the identified model. The difference between an experimental frequency response and the frequency response of the identified model is used to validate the identification in the frequency domain. The identified model corresponding to 30 m/s on a dry road with the average mass has found to be the best robust control design model.

4.5 Design Parameters

The design parameters involved in the control objective are $sh = 1$, $ch = 3$, $ph = 60$, $\rho = 30$, $W_u(z^{-1}) = \frac{1-0.4z^{-1}}{1+1.85z^{-1}+0.855z^{-2}}$ and $W_y(z^{-1}) = 1$.

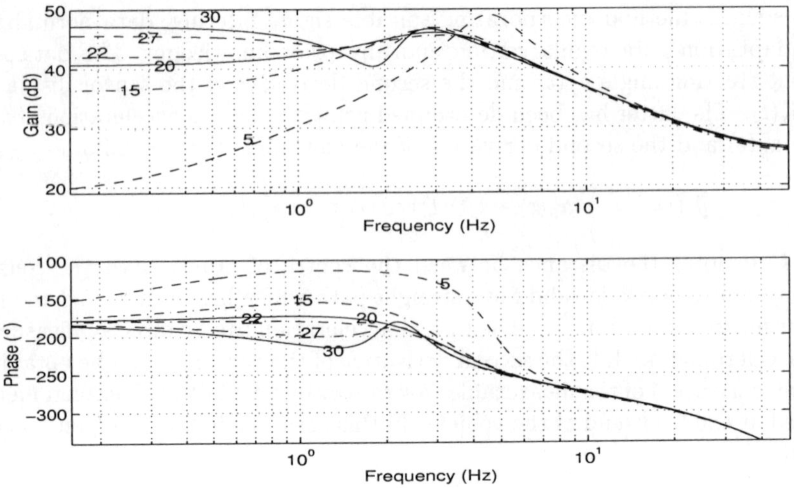

Figure 5: Frequency responses of identified models (dry road and a varying speed)

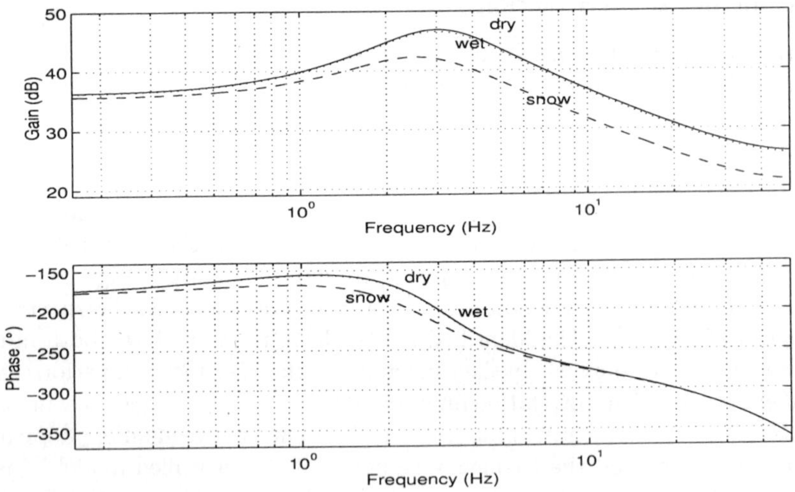

Figure 6: Frequency responses of identified models (constant speed and varying adhesion factor)

The predictor poles are set to 0.5, 0.6, 0.7, 0.8, 0.9 and 0.95. They have been obtained by damping the poles of the control model and adding one auxiliary pole. These design parameters have been specified to achieve the required specifications with adequate shapes of the usual sensitivity functions. The performance of the automatic steering system with the robust controller is shown for the nominal situation and three mismatch conditions, namely

- FC : a fast condition corresponding to a speed of 30 m/s on a dry road (solid lines). It corresponds to the nominal situation.
- OC : an operating condition corresponding to a speed of 18 m/s on a snow road (dashed lines).
- AC : an average condition corresponding to a speed of 15 m/s on a wet road (dash-dotted lines).
- SC : a slow condition corresponding to a speed of 5 m/s on a dry road (dotted lines).

For these conditions the corresponding sensitivity functions are shown in figure 7. The corresponding modulus (MM) and delay margins (DM) as well as the

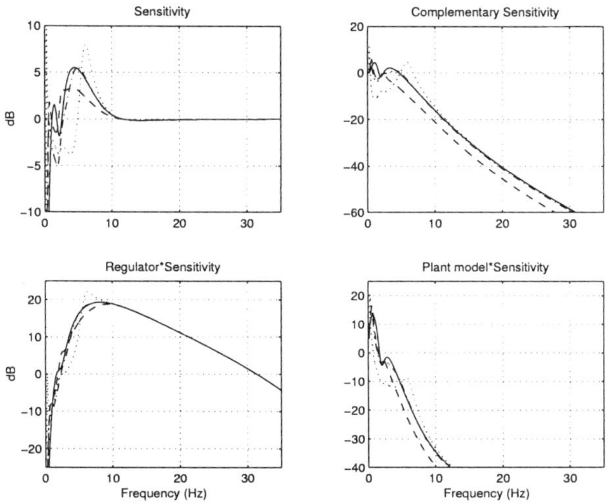

Figure 7: Usual sensitivity functions

attenuation bandwidths (AB) are shown in table 1.

Table 1. Robustness margins and attenuation bandwidths.

Conditions	MM(dB)	DM(sec)	AB(Hz)
FC	-5.51	0.59	1.10
OC	-3.32	1.71	0.57
AC	-5.50	0.61	2.80
SC	-9.61	0.34	0.24

Figures 8 shows the lateral sensor displacements, steering angles, steering angle rates and lateral accelerations for entering into the curved test path. It can be observed that the settling time exhibits large values for low speed condition. Notice that the steering angles and the steering angle rates are more abrupt and exhibit an oscillatory behaviour for large speed conditions and slippery roads. The actuator constraints as well as the passenger comfort specifications have been satisfied with the proposed robust controller. The steady-state values of the steering angle depend on the speed and the road-tyre contact, this corresponds to the understeering behaviour of the BMW520i.
Figure 9 shows the lateral displacements and the steering angles for the slow condition with the robust controller (dotted line) and the underlying nominal controller (solid line). This suggests to think about a cautious parameter adaptation to enhance the performance.

4.6 Cautious Parameter Adaptation

The design parameters involved in the predictive control objective have been specified to meet the above design specifications while ensuring appropriate shaping of the usual sensitivity functions. The control model parameters have been initialized by those of the nominal plant model used to design the underlying robust predictive controller and updated using robust parameter adaptation algorithm incorporating an appropriate signal filtering and normalization, the available prior knowledge, an adequate adaptation gain regularization and a cautious adaptation freezing. The performance of the automatic steering system with suitable parameter adaptation has been demonstrated through severals simulation experiments. For the sake of space limitations, only one experiment will be reported, more specifically,

- The involving road curvature consists in four successive curves and straight lines where the corresponding curvature radius is 400 m. The vehicle slows down from 15 m/s to 5 m/s.

The road curvature, the vehicle speed, the lateral displacement and the steering angle are shown in figure 10. The cautious parameter adaptation has allowed to improve the performance of the automatic steering system. The performance of the robust controller has been enhanced as clearly shown along the distance intervals $[0, 50m)$ and $[50m, 110m)$. Furthermore, the speed deceleration has been accomodated as clearly shown along the distance intervals $[115m, 190m)$ and $[190m, 250m]$. Notice that, that the input-output signals were not sufficiently exciting on straight roads, the parameter adaptation was automatically frozen as the incoming information was likely to improve neither the accurancy of the control model nor its interaction with the underlying control design.

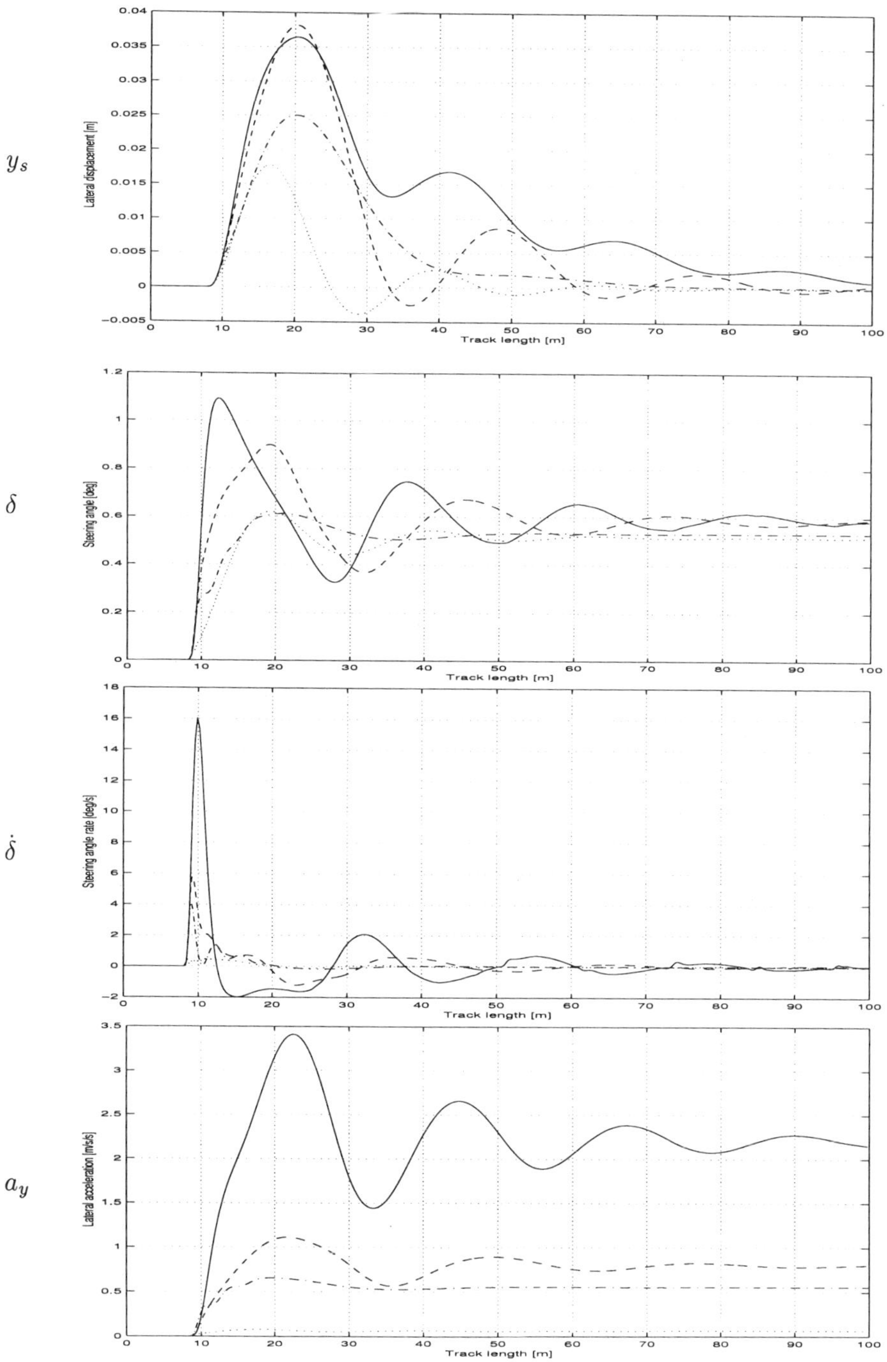

Figure 8: Lateral displacements, steering angles, steering angle rates, and lateral accelerations (curved path)

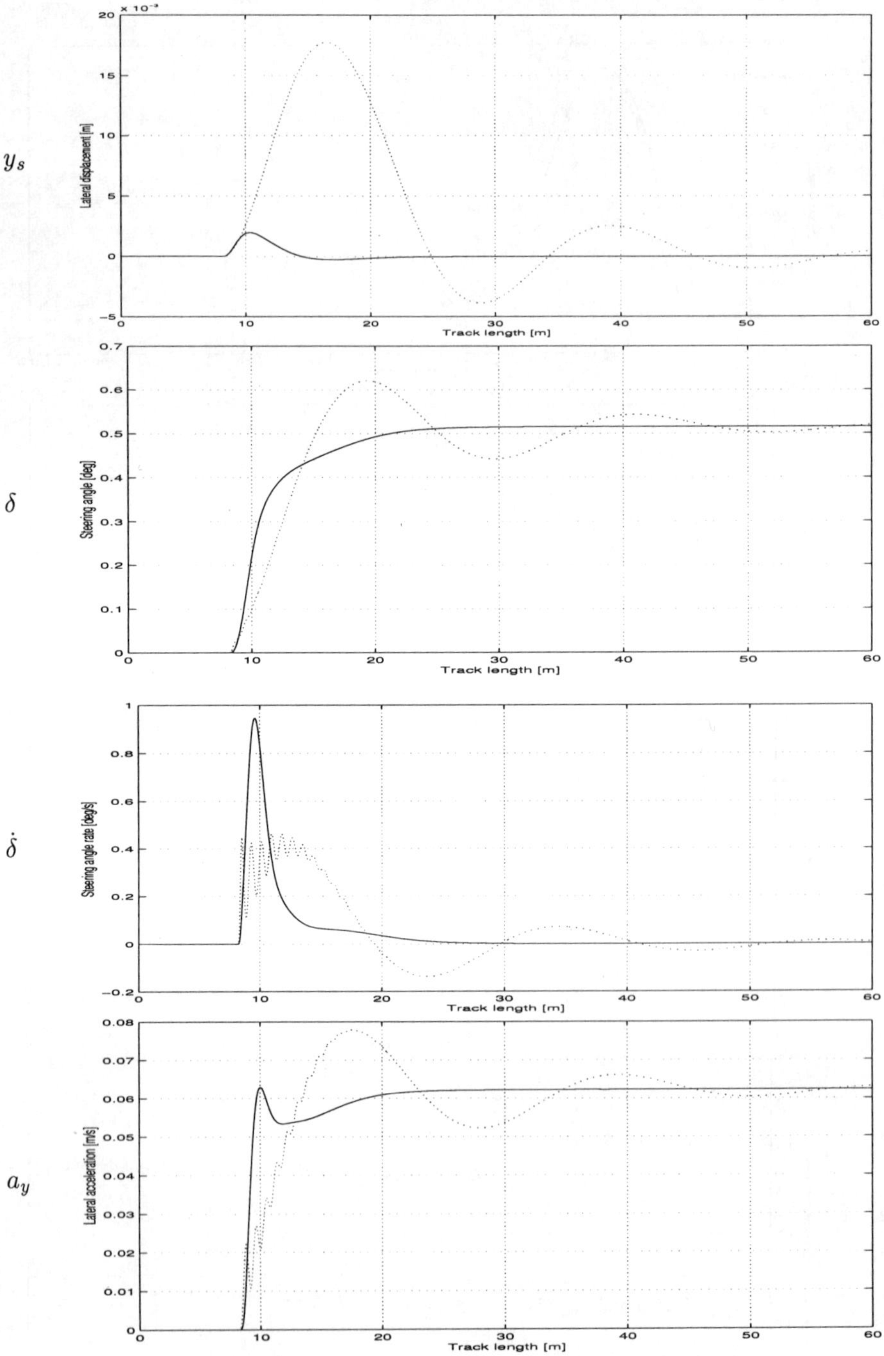

Figure 9: Lateral displacements, steering angles, steering angle rates, and lateral acceleration for nominal and robust controller

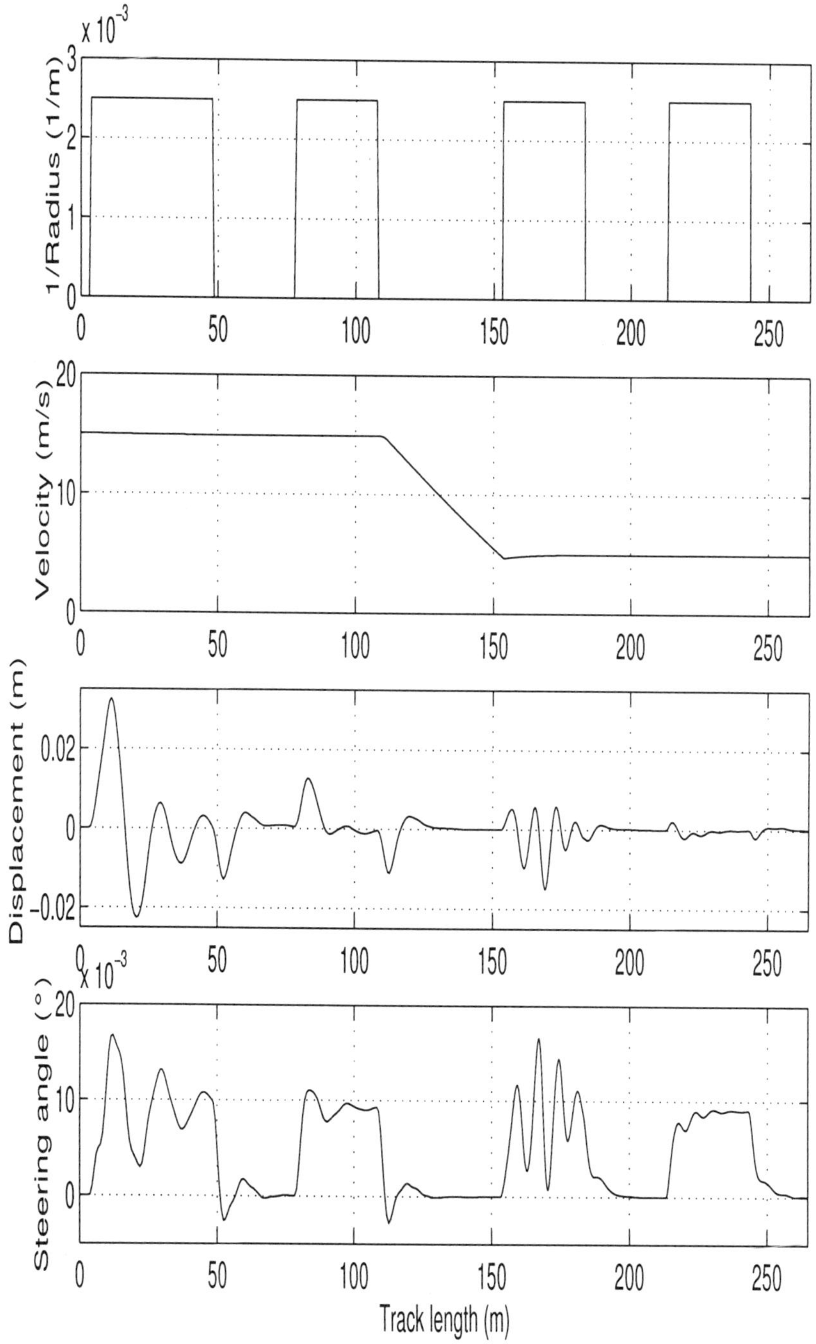

Figure 10: Road curvature, vehicle speed, lateral displacement and steering angle during the simulation with parameter adaptation

5 CONCLUSIONS

The main motivation of this paper was to investigate the applicability of a suitable adaptive robust predictive control approach for the design of an automatic steering system. The standard safety and passenger comfort specifications have been achieved. Three fundamental design features of the proposed control approach are worth to be pointed out, namely offset-free performance, adaptation alertness, stability robustness and implementation simplicity. A great attention has been paid to the identification of the control design model and its parameter adaptation. A comprehensive iterative procedure has been developed to specify the design parameters of the adaptive predictive controller.

REFERENCES

1. Shladover, S. E. (1993). 'Research and development needs for advanced vehicle control systems'. *IEEE Micro*, pp. 11–18.
2. Lee, A. Y. (1992). 'A preview steering autopilot control algorithm for four-wheel-steering passenger vehicles'. *Trans. ASME J. Dyn. Syst., Meas. Control*, **114**, 401–408.
3. Ackermann, J. (1994). 'Robust decoupling, ideal steering dynamic and yaw stabilization of 4ws cars'. *Automatica* **30**(11), 1761–1768.
4. Guldner, J., V.I. Utkin, J. Ackermann and T. Bünte (1995). Sliding mode control for active steering of cars. In 'IFAC-Workshop on AAC'. pp. 57–62. Ascona, Switzerland.
5. Ackermann, J., J. Guldner, W. Sienel, R. Steinhauser and V.I. Utkin (1995). Linear and nonlinear controller design for robust automatic steering. IEEE Trans. on Control Systems Technology, 3 (1), 132–143.
6. Majjad, R. and U. Kiencke (1996*a*). Modelling and simulation of non-linear vehicle dynamics. In 'Proc. MTNS-96'. St. Louis Missouri, U.S.A.
7. Zhou, K., J.C. Doyle and K. Glover (1996). *Robust and Optimal Control*. Prentice-Hall. Englewood Cliffs.
8. Clarke, D. W., C. Mohtadi and P. S. Tuffs (1987). 'Generalized predictive control'. *Automatica*, **23**(1), 23–25.
9. M'Saad, M., R. A. Ramirez-Mendoza and L. Dugard (1996). Automotive control : The rst control approach. In 'IEEE-IFAC CACSD'96'. September, Deadborn, Michigan U.S.A.
10. Majjad, R. and U. Kiencke (1996*b*). Modularer ansatz sur modellierung der kraftfahrzeug dynamik. vdi-tagung : Electronik im kraftfahrzeug. In 'Baden-Baden'.
11. Kiencke, U. (1993). Real estimation of adhesion characteristic between tyres and road. In 'IFAC World Congress'. Sydney, Australia.
12. Kiencke, U. and A. Daiß(1994). Estimation of tyre friction for enhanced abs-systems. In 'Preprints of the AVEG Congress'.
13. Burckhardt, M. (1993). *Fahrwerktechnik: Radschlupfregelsystem.* Vogel Fachbuch.
14. Mitschke, M. (1990). *Dynamik der Kraftfahrzeuge III.* Springer Verlag.
15. M'Saad, M., N. Mueller and R. A. Ramirez-Mendoza (1997). Robust predictive control design for automatic steering. In 'European Control Conference 97'. Brussels, Belgium.

Vehicle System Dynamics Supplement 35 (2001), pp. 181-205

A New CVT Powertrain without Jet Start Behaviour: Analysis of Design, Dynamics and Control

ALEX SERRARENS *, BAS VROEMEN * and FRANS VELDPAUS *

SUMMARY

Due to rotating inertias within the engine and transmission, the response of a vehicle during large engine speed shifts may appear reluctant or even counteracting. Reminiscent of comparable behaviour seen in aircraft jet-propulsion, this phenomenon is also referred to as 'jet-start'. To overcome this behaviour, a CVT powertrain is augmented with a planetary gear set and compact steel flywheel. The new transmission seamlessly combines two contradictory features: the driveability in terms of the pedal-to-wheel response is greatly improved and a large leap towards optimal fuel economy can be made. This is achieved by cruising the vehicle at extremely low engine speeds, using the large ratio-coverage of the CVT. The flywheel acts as a 'peak shaver' during speed shifts: it delivers power during engine acceleration and absorbs kinetic energy during engine decelerations. In this paper, optimal design, system dynamics and control aspects of the new powertrain are discussed.

1 INTRODUCTION

Focus is on the parametric design, dynamics and control of a powertrain incorporating a Continuously Variable Transmission (CVT), a power-splitting parallel stage and a flywheel. The flywheel is present to eliminate the inverse response known as jet-start behaviour, which is common in powertrains that are operated at low engine speeds for fuel efficiency. In general, the term 'powertrain' refers to the entire system which propels the vehicle, arguably extending from energy source to the element interfacing with the surface that provides the reaction force. The powertrain considered can be divided into: petrol as the energy source; the petrol engine as the converter from chemical to mechanical energy; the transmission comprising the torque converter, the variator, possibly the parallel stage and flywheel, the final reduction gear and the differential

*Eindhoven University of Technology, P.O. Box 513, 5600 MB Eindhoven, The Netherlands, E-mail corresponding author: A.F.A.Serrarens@tue.nl

gear; the drive shafts connecting to the wheels; the brakes which can provide negative torques when necessary, and the tires which form the actual interface with the road. For varying purposes, the system boundaries need to be defined differently. More specifically, in the optimization of the functional design of the powertrain the only element considered is the transmission—without even the torque converter—in relation to the vehicle mass to be propelled, whereas the only parts subject to modification are the parallel stage and flywheel. Obviously, for the analysis of the resulting dynamics more needs to be taken into account. The models used for the analysis assume the engine torque to be generated in some way not to be specified, but characterized by its frequency contents only. Furthermore, the brakes are left out of investigation, since the driving part of the powertrain hardly contributes in the braking dynamics. The system considered in the controller design is similar, although torque converter flexibility, and tire slip are omitted intentionally.

2 SYSTEM DESCRIPTION

Figure 1 shows a schematic drawing of what is referred to as 'Zero Inertia' (ZI) powertrain in the sequel. The powertrain comprises a 1.6 *l*. 4 stroke

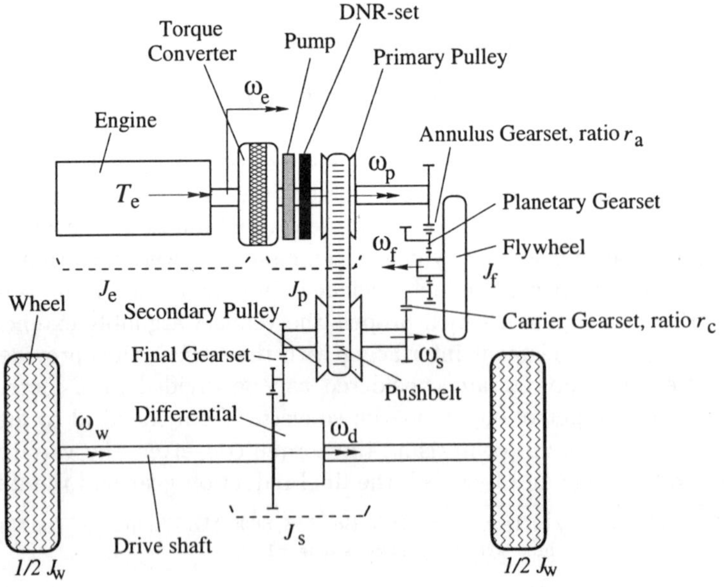

Figure 1: Zero Inertia Powertrain

petrol engine, a commercially available metal pushbelt CVT including auxiliary systems such as a torque converter, a DNR-set, a final reduction gear and a differential. In parallel, a compact steel flywheel is connected to this CVT by a planetary gear set with additional gears but without additional clutches. The planetary gear set operates as a power-split device, deliberately splitting power of the flywheel to and from primary and secondary side whenever the ratio of the CVT is changed. The description of various variables that apply can be found in Appendix A, whereas values of relevant parameters are presented in Appendix D.

The flywheel speed is determined by the primary and secondary speed, *i.e.*

$$\omega_f = \frac{z+1}{r_c}\,\omega_s - \frac{z}{r_a}\,\omega_p, \tag{1}$$

Using the definitions:

$$r = \frac{\omega_s}{\omega_p}, \tag{2}$$

$$\alpha = \frac{z+1}{r_c}; \quad r_{GN} = \frac{z}{\alpha\, r_a}, \tag{3}$$

it follows that the flywheel speed is given by

$$\omega_f = \alpha \left(1 - \frac{r_{GN}}{r}\right) \omega_s \tag{4}$$

Obviously, the flywheel speed becomes zero whenever $r = r_{GN}$. Therefore, r_{GN} is called the geared neutral ratio.

The torque $T_f = J_f\,\dot{\omega}_f$, stemming from the accelerating or decelerating flywheel, is split by the planetary gear set into a torque T_a at the annulus and a torque T_c at the carrier gear, such that

$$T_a = \alpha\, r_{GN}\, J_f\, \dot{\omega}_f; \quad T_c = -\alpha\, J_f\, \dot{\omega}_f, \tag{5}$$

Hence, T_a and T_c are related by

$$T_a = -r_{GN}\, T_c \tag{6}$$

From this last equation the operation of the flywheel unit can be readily explained as follows. If $\dot{\omega}_f$ becomes negative (flywheel decelerates) then the annulus gear *demands* a reaction torque T_a from the primary pulley given by (5) through which the carrier gear can *deliver* a torque T_c to the secondary pulley. This torque T_c is an amplification of T_a if $r_{GN} < 1$. The inverse reasoning holds for an accelerating flywheel. The net torque stemming from the flywheel at the secondary pulley is given by

$$T_n = T_c + \frac{T_a}{r} = -\alpha \left(1 - \frac{r_{GN}}{r}\right) J_f\, \dot{\omega}_f, \tag{7}$$

which is obviously positive for $(r - r_{GN})\dot{\omega}_f < 0$. Taking into account that $\dot{\omega}_s$ is typically an order of magnitude smaller than $\dot{\omega}_f$, it can be seen from equation (4) that the flywheel decelerates as the CVT ratio r decreases, and vice versa. Decreasing the CVT ratio increases the primary speed and thus also the engine speed, resulting in the primary and engine inertias J_p and J_e absorbing part of the torque. Provided the flywheel inertia J_f is sufficient and the CVT speed ratio is manipulated appropriately, then the engine torque T_e may take the acceleration of J_e and J_p on its account, while the flywheel unit delivers the desired net torque T_n as in (7). This is exactly the behaviour necessary to overcome the reluctance in vehicle response whenever large pedal deflections are accompanied by large leaps in engine speed. An optimal design of the geared neutral ratio r_{GN} and J_f is required to let the sizing of the flywheel unit be in concert with its function. This is the subject of the next section.

3 OPTIMIZATION OF SYSTEM DESIGN

In the previous section a number of geometrical and physical parameters of the system were identified. It was decided to add the flywheel unit to the powertrain of a commercially available car. Hence, only the parameters J_f, z, r_a and r_c of the flywheel unit have to be optimized for the chosen car. The value of these parameters determines the functionality of the system, *i.e.* the effectiveness of the flywheel power-boost with a limited CVT ratio range under a small vehicle launch penalty. Clearly, an optimization strategy is required to maximize the boost effectiveness while minimizing the launch penalty. Both criteria therefore need to be quantified in some sense in order to use them in a to-be-minimized objective function.

3.1 Energy functions

Regarding the kinetic energy within the powertrain the optimization criteria can easily be quantified. The purpose of the system is to cancel the disadvantageous effect of the primary inertias using a counteracting flywheel inertia. In terms of energy this means that the total kinetic energy should preferably decrease for decreasing CVT ratio and vice versa. The kinetic energy of the primary inertias (assuming a locked torque converter) is given by

$$E_P = \frac{1}{2}(J_e + J_p)\,\omega_p^2 = \frac{1}{2}(J_e + J_p)\frac{\omega_s^2}{r^2} \tag{8}$$

and for the flywheel by

$$E_F = \frac{1}{2}\,J_f\,\omega_f^2 = \frac{1}{2}\,J_f\,\alpha^2\,r_{GN}^2\left(\frac{1}{r} - \frac{1}{r_{GN}}\right)^2\,\omega_s^2. \tag{9}$$

The kinetic energy of all secondary sided inertias including the vehicle is given by

$$E_S = \frac{1}{2}(J_s + r_f^2\, J_v)\, \omega_s^2. \tag{10}$$

The relative energies of the primary inertias and the flywheel with respect to the secondary inertias follow from

$$e_P = c\,\frac{1}{r^2}; \quad e_F = c\,\gamma\left(\frac{1}{r} - \frac{1}{r_{GN}}\right)^2 \tag{11}$$

where c and γ are defined by

$$c = \frac{J_e + J_p}{J_s + r_f^2\, J_v}; \quad \gamma = \alpha^2\, r_{GN}^2\, \frac{J_f}{J_e + J_p} \tag{12}$$

The constant c is completely determined by parameters of the unmodified powertrain whereas γ depends on the design parameters of the flywheel unit.

Obviously, for a given design of the flywheel unit, the relative energies e_P and e_F and also their sum $e = e_P + e_F$ only depend on the CVT ratio r. For an arbitrary but realistic design these functions are plotted in Figure 2. The

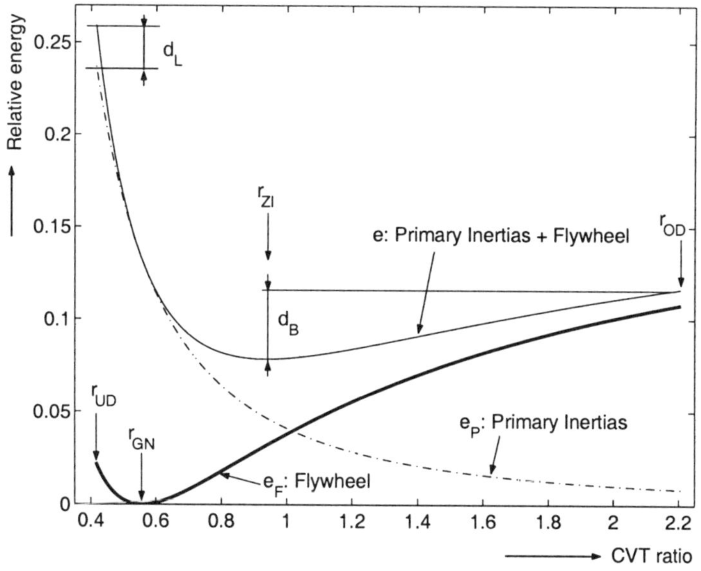

Figure 2: Kinetic Energy Functions

minimum of the total relative energy e occurs for $r = r_{ZI}$, where the so-called zero inertia ratio r_{ZI} is given by:

$$r_{ZI} = \left(1 + \frac{1}{\gamma}\right) r_{GN}, \tag{13}$$

This ratio has to do with the 'zero inertia' situation: when shifting the CVT through this ratio the equivalent inertia felt at the primary side is equal to zero. This remark is in agreement with the observation that the total relative energy does not change when shifting the CVT through this ratio.

In terms of energy, the boost effect of the flywheel can be explained by the decreasing total relative energy for decreasing CVT ratios r as long as $r > r_{ZI}$. For the unmodified powertrain the primary inertias J_e and J_p always require increasing energy for decreasing CVT ratio. Stated otherwise, the released flywheel energy for decreasing CVT ratios will boost the vehicle as long as $r > r_{ZI}$. Therefore, the first variable for the (yet to be defined) objective function F is the so-called boost energy d_B, defined as the difference between the total relative energy for $r = r_{OD}$ and this energy for $r = r_{ZI}$. Using the definition of the zero inertia ratio it is readily seen that

$$d_B = \frac{c}{1+\gamma}\left(\frac{1+\gamma}{r_{OD}} - \frac{\gamma}{r_{GN}}\right)^2 \tag{14}$$

For decreasing CVT ratios r with r between r_{UD} and r_{GN} the increase of the total relative energy $e_P + e_F$ for the powertrain with the flywheel unit is larger than the increase of the relative energy e_P for the unmodified powertrain. Obviously, the largest difference between $e_P + e_F$ and e_P needs to be minimized. This difference, denoted by d_L, occurs for $r = r_{UD}$, so

$$d_L = e_F(r_{UD}) = c\gamma\left(\frac{1}{r_{UD}} - \frac{1}{r_{GN}}\right)^2 \tag{15}$$

This quantity serves as the second variable in the objective function F.

For decreasing CVT ratios r with r between r_{GN} and r_{ZI} the total relative energy for the powertrain with the flywheel unit also increases but less fast than for the unmodified powertrain. Therefore, this part of the curves in Figure 2 is not taken into account in the objective function F.

Summarizing, it can be stated that the boost effectiveness of the flywheel unit is large whenever d_B is large, whereas the penalty on the vehicle launch is small whenever d_L is small.

3.2 Optimization

To optimize the design parameters of the flywheel unit an objective function has to be chosen and the constraints in these parameters have to be formulated. The adopted objective function F is given by

$$F = d_L + \frac{W}{d_B}, \tag{16}$$

where the positive number W weighs the relative importance of the launch penalty d_L over the boost effectiveness d_B. Substitution of the earlier given relations for d_B and d_L yields

$$F = c\gamma\left(\frac{1}{r_{UD}} - \frac{1}{r_{GN}}\right)^2 + \frac{W}{c}(1+\gamma)\left(\frac{1+\gamma}{r_{OD}} - \frac{\gamma}{r_{GN}}\right)^{-2} \tag{17}$$

There are several constraints on the design parameters. From Figure 2 it is clear that r_{GN} must be greater than r_{UD} and that r_{ZI} must be smaller than r_{OD}. Since $r_{GN} < r_{ZI}$ this also implies $r_{GN} < r_{OD}$. With the definition of the zero inertia ratio r_{ZI} and using the fact that $\gamma > 0$, these constraints can be rewritten as

$$r_{GN} > r_{UD}; \quad \gamma(r_{OD} - r_{GN}) > r_{GN}$$

Other constraints have to do with the available space in the engine compartment and the strong desire to minimize the weight of the flywheel unit and the power losses in this unit. The main source of these power losses is the air drag between the rotating flywheel and the flywheel casing. A further elaboration of these desires, combined with the endeavor to make the flywheel as compact as possible, resulted in maximal allowable values $\overline{J}_f$ and $\overline{\omega}_{f,max}$ for, respectively, the moment of inertia and the stationary speed of the flywheel. The values $\overline{J}_f = 0.4$ [kgm^2] and $\overline{\omega}_{f,max} = 750$ [rad/s] turned out to be quite reasonable (see [2]).

To investigate the consequences of the requirement with respect to the maximal stationary flywheel speed it is assumed that the vehicle moves with a known, constant velocity on an horizontal road, experiencing a known load due to air drag and rolling resistance. Then the secondary speed ω_s and the power P_s to drive the vehicle are known. It is assumed that this power P_s is delivered by the engine in an engine operating point on the so-called Peak Efficiency Curve (PEC). More details about this line are given in Section 5. In operating points on the PEC the engine speed (and, since the torque converter is assumed to be locked, also the primary speed ω_p) is a function of the delivered power. Hence, in stationary situations there is a fixed relation between the secondary and the primary speed, *i.e.*, $\omega_p = \omega_{PEC}(\omega_s)$. This relation holds for the unmodified powertrain as well as for the ZI powertrain. Substitution in equation (4) gives the stationary flywheel speed ω_f as a function of ω_s. The requirement $\omega_f \leq \overline{\omega}_{f,max}$ then boils down to

$$\alpha(\omega_s - r_{GN}\,\omega_{PEC}(\omega_s)) \leq \overline{\omega}_{f,max}$$

or, formulated in terms of the parameter α, to $\alpha \leq \alpha_{max}$ with

$$\alpha_{max} = \frac{\overline{\omega}_{f,max}}{\max\limits_{\omega_s}(\omega_s - r_{GN}\,\omega_{PEC}(\omega_s))} \tag{18}$$

#	$\overline{J}_f$	W	d_B	d_L	F	r_{GN}	r_{ZI}
1	0.20	0.020	0.032	0.246	0.866	0.901	1.153
2	0.20	0.004	0.021	0.098	0.287	0.766	1.152
3	0.40	0.020	0.056	0.112	0.448	0.698	0.940
4	0.40	0.004	0.040	0.036	0.136	0.595	0.948
5	0.40	–	0.039	0.029	–	0.577	0.944

Table 1: Optimization results

Combined with the demand $J_f \leq \overline{J}_f$ this results in a constraint for the parameter γ:

$$\gamma \leq \alpha_{max}^2 \, r_{GN}^2 \, \frac{\overline{J}_f}{J_e + J_p} \tag{19}$$

Summarizing, the relevant constraints for the minimization of the objective function F are given by

$$r_{GN} > r_{UD} \tag{20}$$

$$\gamma > 0; \quad \gamma(r_{OD} - r_{GN}) - r_{GN} > 0 \tag{21}$$

$$\gamma(J_e + J_p) - \alpha_{max}^2 \, r_{GN}^2 \, \overline{J}_f \leq 0 \tag{22}$$

The constrained minimization problem is solved for r_{GN} and γ, using the optimization toolbox in MATLAB. Some results for different values of the weight W and the maximal allowable flywheel inertia $\overline{J}_f$ are presented in Table 1.

Solution # 5 in this table stems from an heuristic optimization in which also the gear sizing played an important role. This solution was the starting point for the design of the prototype flywheel unit that is currently assembled and tested. The parameters of this realized design resembles the results of solution # 4 fairly close.

It is inferred from the table that the launch penalty d_L is high whenever the boost effectiveness d_B is high. A larger value for the allowable moment of inertia $\overline{J}_f$ of the flywheel would result in a lower minimum for F but this is not desirable in terms of additional vehicle weight. Applying a lower weight W suppresses the boost effectiveness and consequently also the launch penalty. This illustrates that parameter W can effectively be used to manipulate the trade-off between d_B and d_L.

It can easily be shown that unconstrained minimization of the objective function F does not yield a feasible solution for r_{GN} and γ: the requirement that the derivative of F with respect to r_{GN} is equal to 0 results in a contradiction. Hence, at least one constraint will be active in the constrained minimization of F. For the current minimizations this turns out to be the constraint (22). This implies that for the solution of the constrained problem not $\overline{J}_f$ and $\overline{\omega}_{f,max}$

itself but only the combination $J_f\,\overline{\omega}_{f,max}^2$ is relevant. Much higher maximal speeds than 750 [rad/s] are not desirable as air drag losses within the flywheel housing would increase substantially then. Moreover, the first rotor-dynamical eigenmode of the flywheel will start manifesting.

4 ANALYSIS OF DYNAMICS

To investigate the dynamics of the powertrain with and without the flywheel, it is assumed that the torque converter is locked and can be represented by a linear elastic torsion spring with a relatively high stiffness k_t. The finite stiffness of the drive shafts is lumped in one linear elastic torsion spring with relatively low stiffness k_d. Finally, the tires are modeled as a nonlinear damper, implying that traction is always accompanied by tire slip, be it from deformation or from sliding. This type of tire model is instrumental here because tire slip predominantly determines the damping of powertrain vibrations (see, *e.g.*, [7]).

The system equations in state space representation are given by

$$\dot{x} = f(x) + g(x)\,u \tag{23}$$

with $x = [T_t \;\; \omega_e \;\; T_d \;\; \omega_s \;\; \omega_w \;\; v \;\; r]^T$ the state of the system, v the vehicle velocity, T_t is the torque transmitted by the (locked) torque converter, and T_d is the torque transmitted through the drive shafts. The inputs to the system are the induced engine torque T_e and the CVT shift rate $\dot{r}$, so $u = [T_e \;\; \dot{r}]^T$. Expressions for f and g are given in Appendix B.

A constant state x_0 is called a possible stationary state if there exists a constant input u_0, such that $f(x_0)+g(x_0)\,u_0 = 0$. From the results in Appendix B it follows that such a constant state is determined completely by two of the state variables, for instance by the engine speed ω_{e0} and the CVT ratio r_0.

4.1 Method of analysis

The nonlinear system is linearized around a possible stationary state x_0. The linearized system is given by

$$\dot{\tilde{x}} = A\,\tilde{x} + B\,\tilde{u} \tag{24}$$

where $\tilde{x} = x - x_0$ and $\tilde{u} = u - u_0$. The system matrix A and input matrix B depend on x_0. Expressions for A and B are given in Appendix C.

The linear model is used to investigate some characteristics, like eigenfrequencies and eigenmodes, of the powertrain with and without flywheel. Although it is common practice to linearize around a stationary state, this is not trivial here because in a vehicle powertrain frequently transients occur. One could also consider linearizations around an a priori given state trajectory, resulting in time-dependent matrices A and B. The choice for a stationary state is motivated by the fact that especially the behaviour of the powertrain directly after the start of a transient is of interest here.

The linearizations are performed around several stationary states, characterized by one constant engine speed ω_{e0} and a number of CVT ratios r_0 in the range $[r_{UD}, r_{OD}]$. The characteristic polynomial of the linearized, 7th order system has one pole $s_1 = 0$ in the origin, two real negative poles s_2 and s_3 and four complex poles $s_{4,5} = \beta_1 \pm j\ \Omega_1$ and $s_{6,7} = \beta_2 \pm j\ \Omega_2$ with negative real parts β_1 and β_2. The real poles s_2 and s_3 are determined in the main by the tire properties and the inertia of the vehicle. They are hardly changed if the flywheel unit is added or removed. The complex poles $s_4, ..., s_7$ mainly depend on the inertias of the powertrain components (including the flywheel), on the stiffness k_t of the torque converter and the stiffness k_d of the drive shafts. The imaginary parts (eigenfrequencies) Ω_1 and Ω_2 differ an order of magnitude. The lowest eigenfrequency Ω_1 strongly depends on the stiffness k_d and hardly on k_t whereas Ω_2 is determined in substance by k_t. Therefore, in the following Ω_1 will be referred to as the drive shaft eigenfrequency or drive shaft resonance and Ω_2 as the torque converter eigenfrequency or torque converter resonance.

4.2 Overall dynamics

To illustrate the dynamics of the system with and without flywheel, Figure 3 depicts the transfer functions from the input $\tilde{u}_2 \equiv \dot{\tilde{r}}$ to the vehicle acceleration a_v for a frequency range between 0.1 and 100 [Hz] and for $r_0 \in \{r_{UD},\ r_{GN},\ r_{ZI},$ 1.5, $r_{OD}\}$. Two striking differences become apparent. First, for $r_0 = r_{ZI}$ the amplitude of the transfer function is an order of magnitude smaller than that for the other values of r_0, demonstrating the zero inertia effect. Second, the phase leaps 180 [deg] for $r_0 > r_{ZI}$, indicating that instead of the inverse kick-down response, common to conventional powertrains, the ZI powertrain results in immediate positive accelerations, at least for frequencies up to 2 [Hz]. The amplitude *with* flywheel for these frequencies varies less for different $r_0 > r_{ZI}$ and exceeds the amplitude *without* flywheel by a factor of up to three (in $r_0 = r_{OD}$). For $r_0 = r_{GN}$ there is no difference between the two systems, while for $r_0 = r_{UD}$ the amplitude is larger for the system *with* flywheel than for the system *without*. Since also the phase is undesirable, the behaviour of the ZI powertrain for $r_0 < r_{GN}$ may be characterized as less agreeable.

The second resonance in Figure 3 is far more pronounced and the eigenfrequencies more spread for the ZI driveline. The low (drive shaft) and high (torque converter) resonances will be discussed next.

4.3 The drive shaft resonance

Figure 4 shows some system characteristics at the drive shaft resonance as a function of r_0 for $\omega_{e0} = 100$ [rad/s]. The powertrain without flywheel exhibits a strictly increasing eigenfrequency, whereas the resonance with flywheel has a

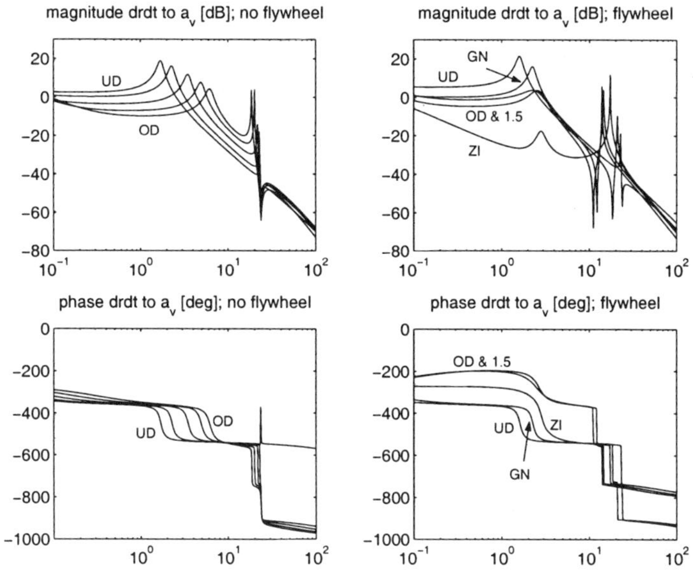

Figure 3: Bode plots without (left) and with (right) flywheel; frequencies in [Hz]

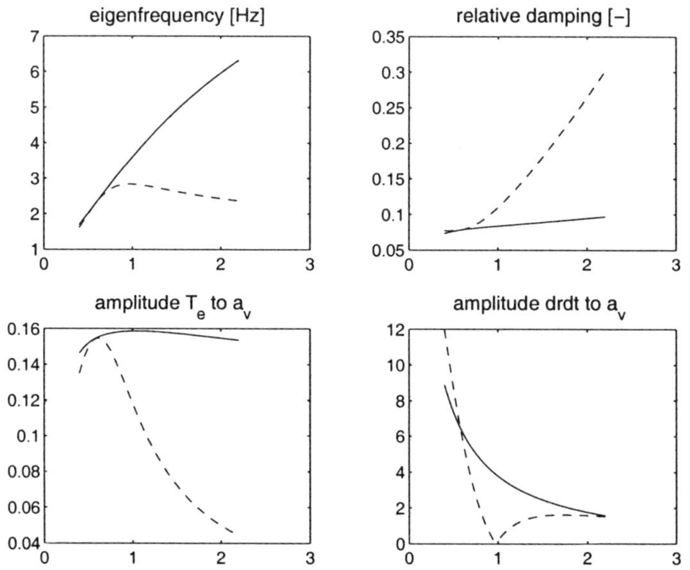

Figure 4: System characteristics vs. r at drive shaft resonance; flywheel is dashed

maximum at $r_0 = r_{ZI}$. The eigenfrequencies for $r_0 < r_{ZI}$ are very similar and are exactly equal for $r_0 = r_{GN}$. The relative damping for $r_0 > r_{GN}$ is substantially larger for the driveline with flywheel. The amplitudes of the transfer functions from the inputs $\tilde{u}_1 \equiv \tilde{T}_e$ and $\tilde{u}_2 \equiv \dot{\tilde{r}}$ to the vehicle acceleration deserve a closer look. The amplitude from $\tilde{T}_e$ to a_v is always smaller (or equal, for $r_0 = r_{GN}$) for the driveline with flywheel, the phase (not shown) is almost equal. The amplitude of the transfer function from $\dot{\tilde{r}}$ to a_v (also shown in Figure 3) is almost zero for $r_0 = r_{ZI}$ while being larger at $r_0 = r_{UD}$ for the system with flywheel than without.

4.4 The torque converter resonance

For the unmodified powertrain the torque converter eigenfrequency decreases with r_0, see Figure 5. This is also true for the ZI powertrain although much

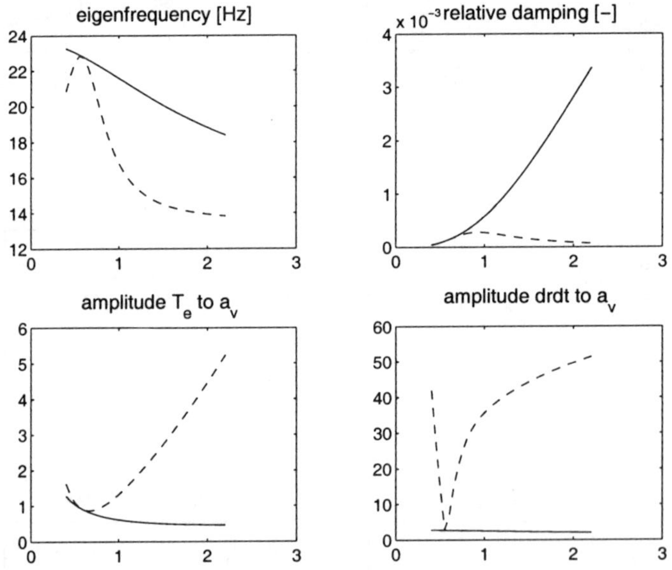

Figure 5: System characteristics vs. r at torque converter resonance; flywheel is dashed

more pronounced from r_{GN} onwards. More disturbingly, the relative damping decreases rather severely for increasing r_0. Also, the amplitudes of the transfer functions from $\tilde{T}_e$ and $\dot{\tilde{r}}$ to a_v are much larger for the ZI driveline. In light of the limited bandwidth of the CVT, this might not be too much of a problem. Furthermore, the transfer from $\tilde{T}_e$ to a_v has something like a notch around the torque converter resonance. Torque variations due to individual combustions (two per crank shaft revolution) are situated roughly between 30 [Hz] (for

$\omega_{e0} = 100$ [rad/s]) and 200 [Hz] (for $\omega_{e0} = 600$ [rad/s]), whereas the bandwidth of the mean value torque varies between 1 [Hz] (for $\omega_{e0} = 100$ [rad/s]) and 3 [Hz] (for $\omega_{e0} > 200$ [rad/s]). Since the eigenfrequencies, as depicted in the upper left plot of Figure 5, can be close to the excitation frequency due to individual combustions, the effect of torque variations at a frequency twice the engine speed needs to be considered in more detail. Figure 6 depicts the amplitude of the transfer function from $\tilde{T}_e$ to a_v evaluated at a frequency of 32 [Hz] ($\omega_{e0} = 100$ [rad/s]). Thanks to the very low relative damping of the torque converter resonance for the flywheel driveline, the amplitude at 32 [Hz] has fallen sharply to a value which is an order of magnitude smaller than the amplitude for the conventional driveline. The downside of this low relative damping is that oscillations, excited by initial conditions unequal to zero, will persist longer and will be of the same order of magnitude as the oscillation at the excitation frequency.

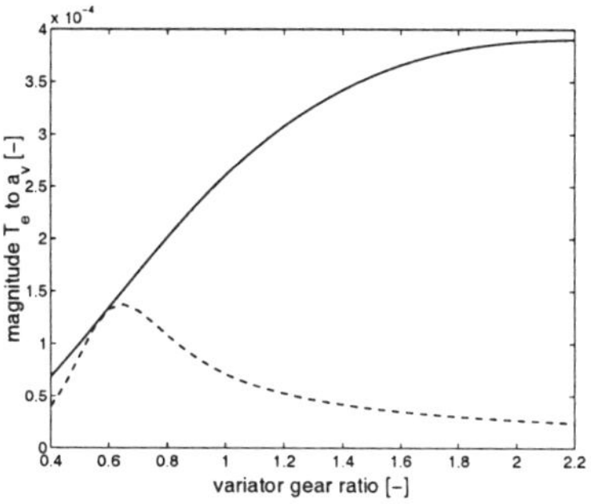

Figure 6: Amplitudes at excitation frequency; flywheel is dashed

5 ANALYSIS OF CONTROL

The ZI powertrain requires a control system to realize the objectives, being minimal fuel consumption and maximal driveability. Here, driveability is seen as a measure for the vehicle response (*e.g.*, the acceleration) directly after a kick down of the drive pedal. For the driver only driveability is of importance in real life traffic situations. The desired behaviour has to be obtained by appropriate control of the engine torque with an electronic air throttle and of the CVT ratio shift speed with an hydraulic system to manipulate the pressures on the CVT pulleys.

5.1 Control hierarchy

The total control of the ZI powertrain consists of three layers, see Figure 7. The

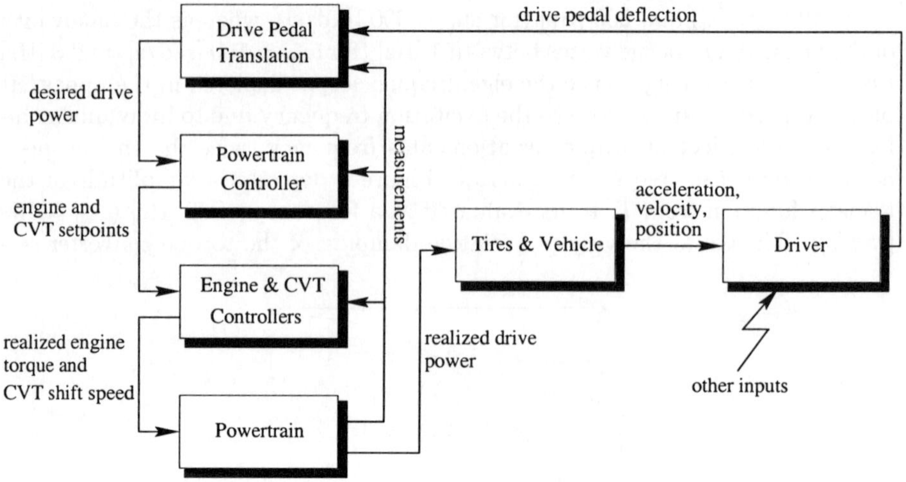

Figure 7: Powertrain Control Hierarchy

first layer is the driver manipulating the drive pedal to control the position, speed and acceleration of the vehicle. The second layer interprets the drive pedal deflections into setpoints for the engine torque and the CVT ratio. An important aspect of this layer is the translation of pedal deflection into desired drive power. The philosophy behind the powertrain controller is that the driver is well able to control the position, speed and acceleration if the error between the desired drive power and the realized drive power is always small. The third layer involves local controllers for the engine throttle and the CVT ratio shift speed. These controllers have to take care that the setpoints, given by the second layer, are realized as good as possible. They are not addressed here in more detail. The local CVT controller is described in [13].

5.2 Controller model

The control problem now is to find setpoints for the engine torque and CVT ratio shift speed such that the fuel economy is maximized (at least in stationary situations) and driveability is maximized for all situations. A reduced order model is proposed to illustrate the difficulties in designing controllers for CVT based powertrains and for the ZI-powertrain in particular. In this model it is assumed that the flexibility of the torque converter spring and the slip between the tires and the road may be neglected. The set of equations of Appendix B then reduces to

$$(J_2(r) + J_e)\,\dot{\omega}_s = r\,T_e + (J_1(r) + J_e)\,\omega_s\,\tfrac{\dot{r}}{r} - r_f\,r^2\,T_d \quad , \qquad (25)$$

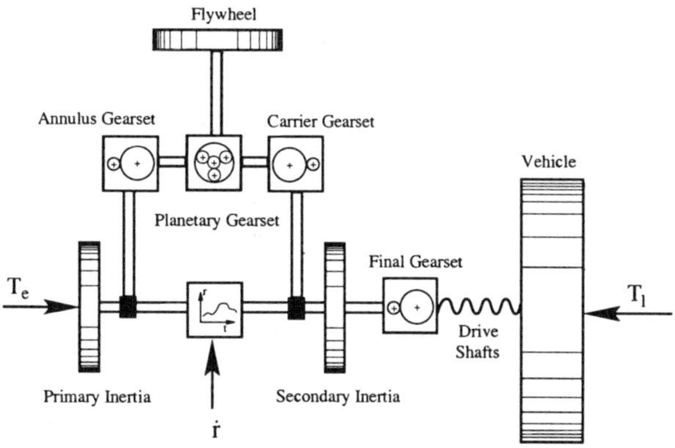

Figure 8: Control Model Zero Inertia Powertrain

$$J_v \, \dot{\omega}_w = T_d - T_l \quad , \tag{26}$$

where $J_v = R_w^2 \, m_v + 2 \, J_w$ is the equivalent moment of inertia of the vehicle plus the wheels, T_d is the torque in the drive shaft and $T_l = T_r + c_v \, R_w^2 \, \omega_w^2$ is the load torque due to rolling resistance and air drag. The meaning of the other symbols is given in Appendix A and Appendix B.

5.3 Assumptions for powertrain controller design

The goal of the powertrain controller is to realize a drive shaft torque T_d that follows a time-dependent torque $T_{d,d}$ as close as possible, using as little fuel as possible. The controller has to cope with a number of constraints, being

- the CVT ratio is bounded by r_{UD} and r_{OD}. Especially the constraint $r \geq r_{UD}$ is very restrictive
- the engine speed is bounded by $\omega_{e,min}$ and $\omega_{e,max}$ where $\omega_{e,max}$ is the speed at maximal engine power
- for each engine speed ω_e the engine torque is at the utmost equal to the wide open throttle torque $T_{WOT}(\omega_e)$.

Taking these constraints into account, the desired drive shaft torque $T_{d,d}$ is derived from the drive pedal deflection ρ (scaled between 0 and 1 with $\rho = 1$ at maximal deflection) as:

$$\dot{P}_{d,d} = \kappa(P(\rho) - P_{d,d}); \quad T_{d,d} = \begin{cases} \rho \, \frac{T_{WOT}(\hat{\omega}_e)}{r_f \, r_{UD}} & \text{for } \hat{\omega}_s \leq r_{UD} \, \omega_{e,max} \\ \frac{P_{d,d}}{r_f \, \hat{\omega}_s} & \text{for } \hat{\omega}_s > r_{UD} \, \omega_{e,max} \end{cases} \tag{27}$$

where $P_{d,d}$ is the desired power through the drive shaft, κ can be seen as a 'pedal-to-wheel stiffness', $P(\rho)$ is a yet to be chosen, strictly increasing function of the drive pedal deflection ρ and, finally, $\hat{\omega}_e$ and $\hat{\omega}_s$ are the measured engine and secondary speed, respectively. Thus, dependent on whether or not the vehicle exceeds a certain minimum speed, the drive pedal deflection is translated into a desired torque, either directly or indirectly via a power interpretation. The choice for the drive shaft torque in this translation is argued as follows. The time scale of varying vehicle loads in daily life traffic is generally large enough for the driver to take safe and appropriate actions. This fact, combined with the knowledge that common drivers are well able to control second order dynamics (*e.g.* see [6]), makes it reasonable to assume that the driver feels comfortable with the task of controlling the longitudinal *vehicle* dynamics, and less comfortable (though maybe more sportive) with *powertrain* dynamics. Unlike in [9], vehicle speed control therefore is not seen as a task of the powertrain controller unless explicitly demanded for by the driver by means of an enabled cruise controller. The drive shaft torque T_d is exactly the quantity the driver needs to control the slowly varying low order vehicle dynamics (26) towards a desired vehicle speed or a relative distance to other vehicles in traffic. Although other solutions may apply, the translation of the pedal deflection ρ into $T_{d,d}$ via a power interpretation as in the second case of equation (27), certifies that maximal engine power $T_{WOT}(\omega_{e,max})\,\omega_{e,max}$ is always translated into a maximal drive shaft torque for all secondary speeds. Many authors ([10], [11] and [5]) agree in translating the drive pedal deflection into a desired drive shaft torque, but none of them really gives an actual solution. In [12] the power interpretation is also proposed since investigation of numerous experiments showed power to have the best correlation with pedal deflection. For both cases in (27) it is asserted that longitudinal driveability is 'high' whenever the difference between T_d and $T_{d,d}$ is 'small' for every time t.

5.4 Control of driveability vs. fuel economy

The powertrain controller can use the engine torque T_e and the CVT ratio shift speed $\dot{r}$ to control T_d. Choosing setpoints $T_{e,d}$ for T_e and $\dot{r}_d$ for $\dot{r}$ is far from trivial since both inputs can contribute to driveability and fuel economy (see also [4] and [8]).

Compared to CVT powertrains without an assisting flywheel, the driveability of the ZI powertrain can be improved considerably. Maximal fuel economy is obtained if every stationary engine power is delivered in an operating point with minimum brake specific fuel consumption. The collection of all such points unveils the Peak Efficiency Curve (PEC) as shown in Figure 9. The Optimal Operating Line (OOL) lines up all operating points minimizing the fuel consumption for stationary throttle positions. Using an electronic air throttle the fuel consumption can be minimized by exactly tracking the PEC ([1]). With a measured engine speed $\hat{\omega}_e$, the PEC is tracked as good as possible if the engine

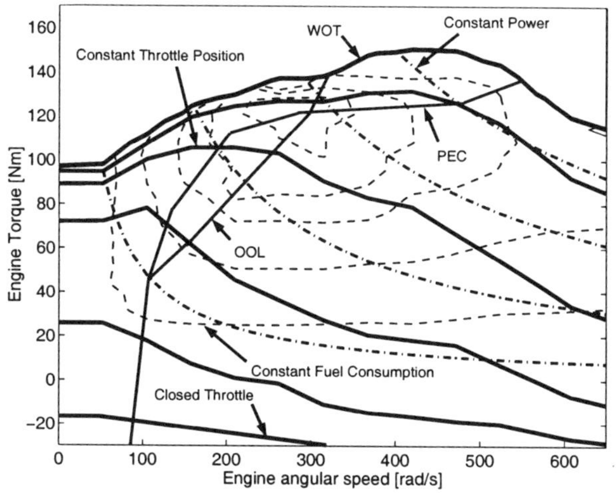

Figure 9: Minimal fuel consumption lines in the engine map

torque T_e is equal to the setpoint $T_{e,d}$ with

$$T_{e,d} = \min(T_{PEC}(\hat{\omega}_e), r_f\, r_{OD}\, T_{d,d}), \tag{28}$$

This setpoint is below the PEC whenever stationary torques are being delivered at $r = r_{OD}$, *i.e.* at OverDrive. Rearranging equation (25) with appropriate substitutions, the setpoint for the CVT speed ratio is then found to be:

$$\dot{r}_d = \frac{(J_2(r_d) + J_e)\, r_d\, \hat{\dot{\omega}}_s + r_f\, r_d^3\, T_{d,d} - r_d^2\, T_{e,d}}{(J_1(r_d) + J_e)\, \hat{\omega}_s}, \tag{29}$$

where $\hat{\dot{\omega}}_s$ is a reconstruction of the secondary pulley acceleration.

For the ZI powertrain, $J_1(r) + J_e$ is negative for ratios between r_{OD} and r_{ZI}. Within this range, the control law (29) generates a shifting towards lower ratios for torque demands higher than can actually be delivered by the engine. This downshifting brings the engine power towards a higher level by increasing the engine speed and, more important, during the shifting itself it exactly fills up the torque gap between T_e and T_d in equation (25). For ratios r_d approaching r_{ZI}, control law (29) can not be used because $J_1(r_d) + J_e$ then approaches zero. Controlling the CVT ratio for the ZI powertrain then raises the same type of difficulties as for the unmodified powertrain. For unmodified CVT powertrains $J_1(r) + J_e = J_p + J_e$ is always positive and control law (29) would result for all possible CVT ratios in an unwanted upshift for torque demands higher than actually can be delivered by the engine. In [3], exact feedback linearization of the powertrain dynamics is used to develop a nonlinear compensator for the control of a desired drive torque. Although the resulting controller is time-optimal in the linearized system coordinates, driveability in terms of realized drive torque could not be guaranteed. In practice, well behaving controllers are developed that relax the ratio shift speed to assure a strictly increasing though less time-optimal torque response in the drive shaft. Deviations from the PEC up to the WOT line will increase the driveability although it is typically traded for fuel economy. The solution chosen here is to saturate $J_1(r_d) + J_e$ in (29) at some value smaller than zero. This will indeed relax the shift speed. Within the operation envelope of engine and vehicle, driveability results are fair. Note that the equivalent primary inertia $J_1(r) + J_e$ will become positive for r between r_{ZI} and r_{GN} although it is still less than for a conventional CVT powertrain. Ratio's below r_{GN} only occur during vehicle launch and during fast accelerations from low vehicle speeds. As discussed in Section 3, the performance penalty for this case is minimized in the current design of the flywheel unit and is, anyhow, for the larger part determined by the primary inertias (d_L in Figure 2 is small with respect to e_P).

5.5 Stability

The stability of the closed loop system with the controllers (28) and (29) is not yet determined in a closed form. A large number of simulations, including simulations with realistic disturbances from the engine and the wheels, did not exhibit unstable behaviour. Non-linear control solutions for $\dot{r}_d$ backboned by an asymptotic convergence law for the engine speed towards a final value assure stability but generally constitute a less predictable driveability. Uncovering the stability properties of the closed loop internal dynamics remains a topic for future research.

5.6 Simulation results

In this section simulation results are shown, using three different strategies for determining the setpoint for the engine torque. The simulations are performed with model (23), augmented with non-linear second order CVT dynamics and slightly less than critically damped second order engine dynamics. Some of the results are presented in Figure 10 and Figure 11. The first strategy, indicated by 'a', is given by equation (28) and tracks the PEC also in transients. Strategy 'b' can deviate from the PEC to a torque level corresponding to:

$$T_{e,d} = \min(T_{PEC}(\omega_{e,d}), r_f \, r_{OD} \, T_{d,d}), \tag{30}$$

where $\omega_{e,d}$ is the engine speed in the operating point on the PEC, corresponding to $P_{d,d}$. The third strategy, indicated by 'c', controls the engine torque along the WOT line as long as the engine power is smaller than $P_{d,d}$. The control strategy then becomes:

$$T_{e,d} = \min\left(T_{WOT}, \max\left(\frac{P_{d,d}}{\hat{\omega}_e}, \min(T_{PEC}(\hat{\omega}_e), r_f\, r_{OD}\, T_{d,d})\right)\right) \tag{31}$$

Strategies similar to b and c, also referred to as "Off the Beaten Track", were proposed earlier in [12] for the so-called economy and sport mode for CVT-equipped passenger vehicles.

Excursions of T_e using these three strategies can be seen in the lower right plot in Figure 10. The upper left plot shows the desired and realized power for the three strategies, originating from an arbitrarily chosen deflection of the drive pedal travel as a function of time t. The upper right plot shows a fragment thereof and the lower right plot gives the same fragment but now displayed in the engine map. The cumulative fuel consumption for the three strategies is presented in the lower left plot, whereas the engine speed, flywheel speed and vehicle speed and acceleration can be seen in Figure 11. From these plots the following observations and conclusions evolve:

- The driveability is about the same for all three strategies. The differences in engine torque are compensated by the ratio shifting law (29), whereas the fuel consumption is slightly higher (ca 0.9 %) for strategy c.
- There is a large difference between the flywheel and engine speed transients for strategy c and those for the strategies a and b.
- As explained in paragraph 5.4, inertial torque compensation is possible only for $r > r_{ZI}$. For $r < r_{ZI}$, strategy c might alleviate a slight reluctance in performance, with a small fuel economy penalty.
- The initial operating point of the engine is not at all on PEC. This is caused by the CVT ratio constraint $r \leq r_{OD}$.
- The lower right plot in Figure 11 shows that the response in vehicle accelerations is fairly smooth though immediate.
- The oscillations of what was referred to as torque converter resonance in Section 4 seem acceptably low in magnitude.
- To obtain these smooth results the reconstruction of $\hat{\omega}_s$, used in equation (29), was filtered using a 1 [rad/s] bandwidth first order filter. This bandwidth is well below the drive shaft resonance, assuring that this resonance is not amplified by the CVT actuation.

6 CONCLUSIONS

A system description of a new CVT powertrain with an extra flywheel is presented. This flywheel fills the driveability gap of conventional powertrains that are controlled for maximal fuel economy. Dynamic analysis showed that the

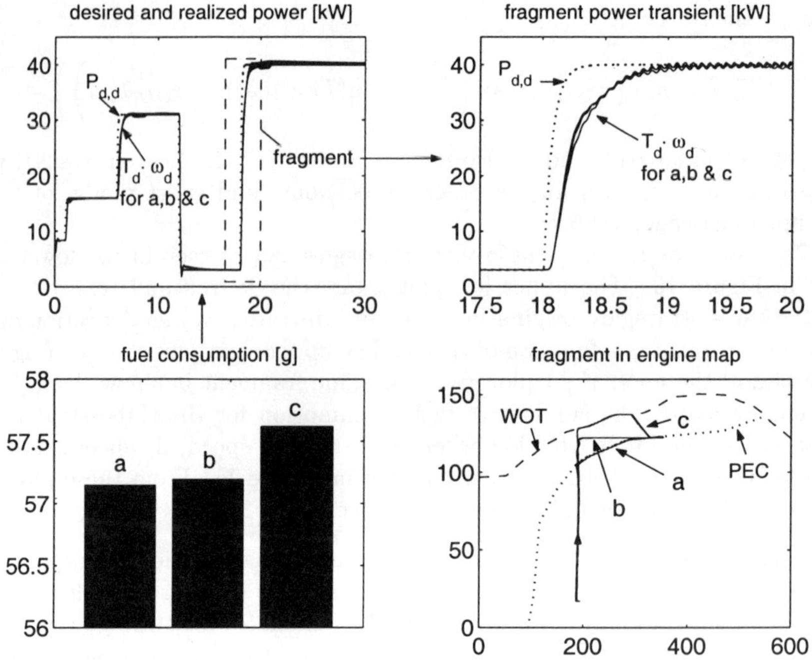

Figure 10: Simulation Results

leading resonance modes of the new powertrain are quite different. Oscillations induced by the torque converter may be more apparent though still small in magnitude. Oscillations induced by the drive shafts receive extra (virtual) damping up to a factor three by the additional flywheel. A powertrain controller was derived, generating setpoints for engine and CVT from the course of the drive pedal deflection. This controller succeeds in controlling the engine speed at the lowest possible values for stationary situations while delivering smooth though immediate acceleration responses of the vehicle. The system has already been tested extensively and successfully on a test rig. In the near future, the objective and subjective response of a real vehicle with the new powertrain will be evaluated.

ACKNOWLEDGEMENTS

This study is part of the EcoDrive project, a joint project of Van Doorne's Transmissie (VDT), Netherlands Organization for Applied Scientific Research (TNO) and

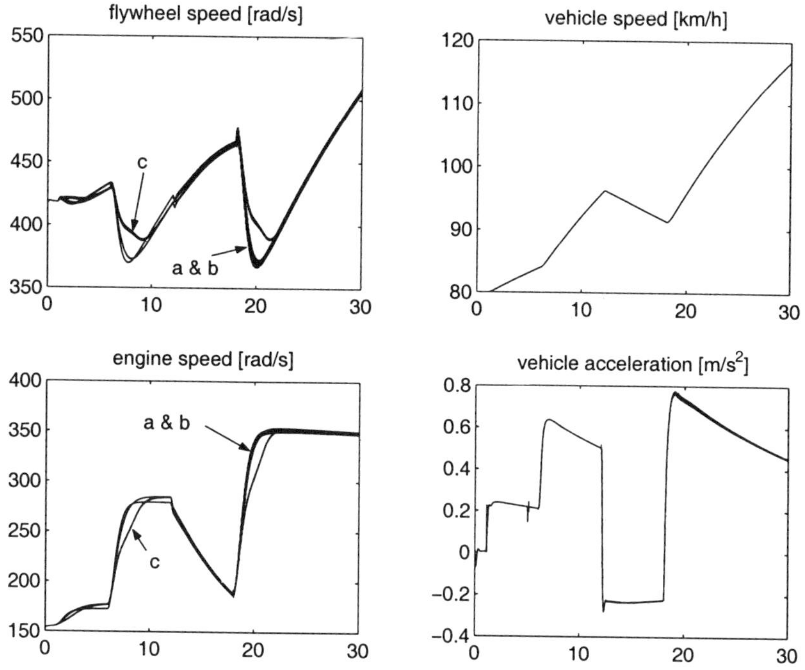

Figure 11: Vehicle response

the Technical University Eindhoven (TUE). The project is subsidized by the Dutch governmental program E.E.T. (Economy, Ecology and Technology).

REFERENCES

1. Chan, C., Yang, D., Volz, T., Breitweiser, D., Jamzadeh, F.S., Frank, A. and Omitsu, T., "System Design and Control Considerations of Automotive Continuously Variable Transmissions, " *SAE Technical Paper 850569*, 1985.
2. Druten, R.M. van, and Kok, D.B., "Design Optimization of a Compact Flywheel System for Passenger Cars, " *VDI Berichte Nr. 1459*, 1999.
3. Guzella, L. and Schmid, A.M., "Feedback Linearization of Spark-Ignition Engines with Continuously Variable Transmissions, " *IEEE Trans. Ctl. Sys. Techn.*, vol. 3, no. 1, 1995.
4. Liu, S., Stefanopoulou, A.G., and Paden, B.E., "Effects of control structure on performance for an automotive powertrain with a continuously variable transmission, " *5th Int'l Symposium on Advanced Vehicle Control*, Ann Arbor, August 2000.

5. Mayer, T., and Schroeder D., "Operating Modes and Control Aspects for a Special Hybrid Drivetrain, " *1st IFAC workshop on Advances in Automotive Control*, 1995.
6. McRuer, D., and Krendel E., "Dynamic response of human operators, " *Columbus Wright Air Development Center*, 1957.
7. Mitschke, M., "Dynamik der Kraftfahrzeuge–Band B: Schwingungen, " *Springer*, 1997.
8. Pfiffner, R., "Steuerstrategien von CVT-Getrieben: Übersicht und Ausblick, " *VDI Berichte Nr. 1467*, 1999
9. Schmid, A., Dietrich, P., Ginsburg, S., and Geering, H.P., "Controlling a CVT-Equipped Hybrid Car, " *SAE Technical Paper 950492*, 1995.
10. Shafai E., Schmid, A., and Geering H.P., "Torque Pedal for a Car with a Continuously Variable Transmission, " *SAE Technical Paper 941010*, 1994.
11. Streib, H-M., and Leonard, R., "Hierarchical control strategy for powertrain functions, " *C389/342, IMechE*, 1992.
12. Vahabzadeh, H, and Linzell, S.M., "Modelling, Simulation, and Control Implementation for a Split-Torque, Geared Neutral, Infinitely Variable Transmission, " *SAE Technical Paper 910409*, 1991.
13. Vroemen, B.G., Serrarens, A.F.A., and Veldpaus, F.E., "CVT Control: a Hierarchical Approach, " *5th Int'l Symposium on Advanced Vehicle Control*, Ann Arbor, August 2000.
14. Vroemen, B.G., and Veldpaus, F.E., "Control of a CVT in a Flywheel Assisted Driveline, " *FISITA World Automotive Congress*, Seoul, 2000.

APPENDIX A: NOMENCLATURE

variable	description
v_v	vehicle speed
ω_d	angular speed of the differential gear
ω_e	angular speed of the engine
ω_f	angular speed of the flywheel
ω_p	angular speed of the primary pulley
ω_s	angular speed of the secondary pulley
ω_w	angular speed of the wheels
$\hat{\omega}_e$	reconstruction angular speed of the engine
$\hat{\omega}_s$	reconstruction angular speed of the secondary pulley
J_e	moment of inertia of rotating engine components
J_f	moment of inertia of the flywheel
J_p	total moment of inertia primary sided transmission components
J_s	total moment of inertia secondary sided transmission components
J_v	equivalent moment of inertia of the vehicle plus wheels
J_w	moment of inertia of two wheels
J_1	primary sided equivalent moment of inertia
J_2	secondary sided equivalent moment of inertia
r	CVT speed ratio
r_d	desired CVT speed ratio
r_a	annulus gear ratio
r_c	carrier gear ratio
r_f	final drive gear ratio
r_{UD}	lowest CVT speed ratio (underdrive)
r_{OD}	highest CVT speed ratio (overdrive)
r_{GN}	CVT ratio where ω_f is zero (geared neutral)
r_{ZI}	CVT ratio at which J_1 is zero (zero inertia)
z	characteristic ratio of planetary gear set
α	cluster of parallel stage gear ratios
T_a	reaction torque of annulus at primary pulley
T_c	reaction torque of carrier at secondary pulley
T_d	summed torque within drive shafts
T_e	mean value brake engine torque
T_n	net torque from flywheel unit at secondary pulley
T_t	torque within (locked) torque converter
b_w	tire-road slip coefficient
c_v	air drag coefficient
k_d	drive shaft stiffness
k_t	torque converter spring stiffness
m_v	vehicle mass
R_w	tire roll radius
T_r	rolling resistance torque
P	desired stationary power through the drive shafts
$P_{d,d}$	desired (instationary) power through the drive shafts
$T_{e,d}$	desired mean value brake engine torque
T_{PEC}	engine torque on Peak Efficiency Line
T_{WOT}	engine torque for Wide Open Throttle
γ	relative flywheel inertia
κ	pedal-stiffness coefficient
ρ	pedal deflection

APPENDIX B: SYSTEM EQUATIONS

The nonlinear system equations are given by

$$\dot{T}_t = k_t\left(\omega_e - \frac{\omega_s}{r}\right) \tag{32}$$

$$\dot{\omega}_e = \frac{1}{J_e}(u_1 - T_t); \quad u_1 = T_e \tag{33}$$

$$\dot{T}_d = k_d(r_f\omega_s - \omega_w) \tag{34}$$

$$\dot{\omega}_s = \frac{r^2}{J_2(r)}\left(\frac{T_t}{r} - r_f\,T_d + \frac{u_2}{r^3}\,\omega_s\,J_1(r)\right) \tag{35}$$

$$\dot{\omega}_w = \frac{1}{J_w}\left(T_d - R_w\,b_w\left(1 - \frac{v_v}{R_w\,\omega_w}\right) - T_r\right) \tag{36}$$

$$\dot{v}_v = \frac{1}{m_v}\left(b_w\left(1 - \frac{v_v}{R_w\,\omega_w}\right) - c_v\,v_v^2\right) \tag{37}$$

$$\dot{r} = u_2 \tag{38}$$

where $J_1 = r_{GN}(r_{GN} - r)\,\alpha^2\,J_f + J_p$ and $J_2 = (r_{GN} - r)^2\,\alpha^2\,J_f + J_p + r^2\,J_s$.

APPENDIX C: MATRICES OF THE LINEARIZED MODEL

$$A = \begin{bmatrix} 0 & A_{12} & 0 & A_{14} & 0 & 0 & A_{17} \\ A_{21} & 0 & 0 & 0 & 0 & 0 & 0 \\ 0 & 0 & 0 & A_{34} & A_{35} & 0 & 0 \\ A_{41} & 0 & A_{43} & 0 & 0 & 0 & A_{47} \\ 0 & 0 & A_{53} & 0 & A_{55} & A_{56} & 0 \\ 0 & 0 & 0 & 0 & A_{65} & A_{66} & 0 \\ 0 & 0 & 0 & 0 & 0 & 0 & 0 \end{bmatrix} \tag{39}$$

where

$$A_{12} = k_t \tag{40}$$

$$A_{14} = \frac{-k_t}{r_0} \tag{41}$$

$$A_{17} = \frac{k_t\,\omega_{s0}}{r_0^2} \tag{42}$$

$$A_{21} = \frac{-1}{J_e} \tag{43}$$

$$A_{34} = r_f\,k_d \tag{44}$$

$$A_{35} = -k_d \tag{45}$$

$$A_{41} = \frac{r_0}{J_2(r_0)} \tag{46}$$

$$A_{43} = \frac{-r_f\, r_0^2}{J_2(r_0)} \tag{47}$$

$$A_{47} = \frac{-T_{t0}}{J_2(r_0)} \tag{48}$$

$$A_{53} = \frac{1}{J_w} \tag{49}$$

$$A_{55} = -\frac{b_w\, v_{v0}}{J_w\, \omega_{w0}^2} \tag{50}$$

$$A_{56} = \frac{b_w}{J_w\, \omega_{w0}} \tag{51}$$

$$A_{65} = \frac{b_w\, v_{v0}}{m_v\, R_w\, \omega_{w0}^2} \tag{52}$$

$$A_{66} = -\frac{b_w + 2c_v\, R_w\, v_{v0}\, \omega_{w0}}{m_v\, R_w\, \omega_{w0}} \tag{53}$$

$$B = \begin{bmatrix} 0 & 0 \\ B_{21} & 0 \\ 0 & 0 \\ 0 & B_{42} \\ 0 & 0 \\ 0 & 0 \\ 0 & 1 \end{bmatrix} \tag{54}$$

where

$$B_{21} = \frac{1}{J_e} \tag{55}$$

$$B_{42} = \frac{J_1(r_0)\, \omega_{s0}}{J_2(r_0)\, r_0} \tag{56}$$

APPENDIX D: VEHICLE PARAMETERS

parameter	unit	value	parameter	unit	value
J_e	kg m^2	0.165	R_w	m	0.307
J_p	kg m^2	0.079	r_f	-	0.2127
J_s	kg m^2	0.034	r_{UD}	-	0.416
J_f	kg m^2	0.40	r_{OD}	-	2.210
J_w	kg m^2	1.70	r_{GN}	-	0.577
m_v	kg	1368	r_{ZI}	-	0.944
k_t	Nm/rad	1182	κ	s^{-1}	15
k_d	Nm/rad	3120	T_r	Nm	55
b_w	N	$5.36 \cdot 10^4$	c_v	kg/m	0.367

Vehicle System Dynamics Supplement 35 (2001), pp. 207-222

Optimization of Automatic Gearshifting

ALI HAJ-FRAJ and FRIEDRICH PFEIFFER*

SUMMARY

Based on a verified mechanical model for a powertrain with automatic transmission, an approach for the optimization of gear shift operations in automatic transmissions is presented. The control data for the clutch pressure in the gear box and the engine load-reduction are used as optimization parameters. The objective of the optimization is to determine an optimal set of control data, so that the performance of the automatic transmission with respect to passengers comfort and gear shift duration is improved. For the description of the comfort, two optimization criteria are formulated. Furthermore, a third criterion is considered for the the evaluation of the gear shift duration. The nonlinear optimization problem is solved with a sequential quadratic programming algorithm. The results show a significant improvement in terms of the optimization criteria.

Keywords: power train dynamics, automatic transmission, gear shift process, comfort, optimization, pareto-optimal solution.

1 INTRODUCTION

Nowadays, electronic components represent an essential element in the car. Their use in combination with optimal control applications makes it possible to fulfill the increasing requirements for more comfort, less fuel consumption and higher power performance. Therefore passenger cars have become very complicated mechatronic systems, where many power train components have to be controlled simultaneously. Figure 1 shows a part of the electronic car management with a Controller Area Network (CAN), which is responsible for the data exchange between the different electronic devices in the car.

In this paper we will concentrate on the optimization of the gear shift process, which is generally carried out by controlling oil-immersed clutches and brakes. The main criteria for the evaluation of the shift quality are the duration of the gear shift process and some measures for the emerging acceleration change.

*Institute B of Mechanics, Technische Universität München 85748 Garching Germany

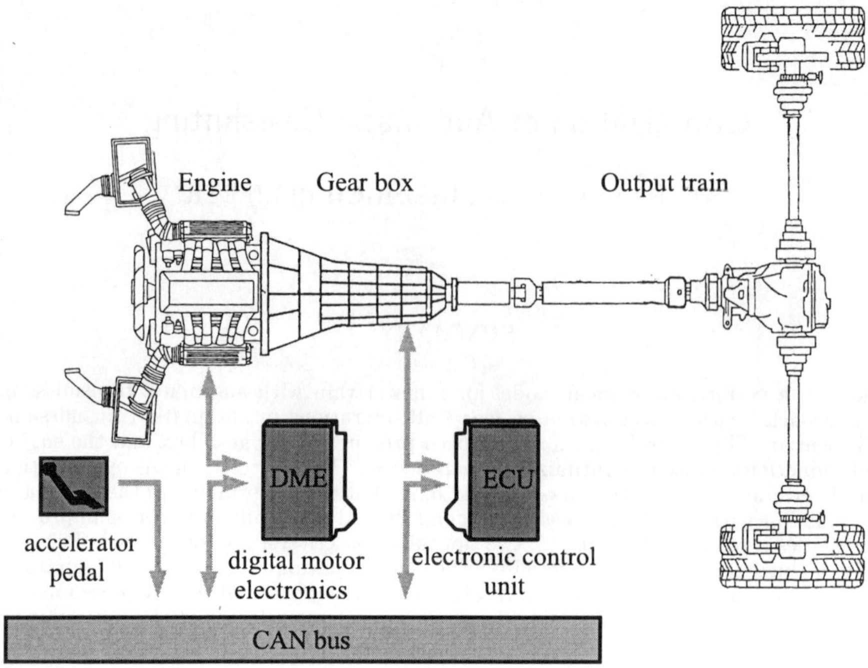

Figure 1: Electronic control of the engine and the automatic transmission

During a gear shift operation two clutches are operating simultaneously, one clutch engaging and the other one disengaging. The clutch pressure is determined by the electronic control unit (ECU) and applied by the hydraulic device of the gear box. At the same time the engine torque is reduced by the digital motor electronics (DME) by retarding the ignition timing or adjusting the opening of the throttle.

In [2] and [6] the fundamentals of powershifting in automatic transmissions using a simple model are investigated. An efficient method for detailed modeling of power trains with automatic transmissions as multibody systems with time-varying structure is presented in [4]. An approach for the evaluation of the passengers comfort during the gear shift process was developed in [3]. In this paper, we propose a method for a model based optimization of gear shift operations in automatic transmissions.

2 MECHANICAL MODEL

A drive train with automatic transmission in general consists of five main components: engine, torque converter, gear box, output train and vehicle. Each component of the drive train may be considered as a rigid multibody subsystem. The rigid bodies are connected to each other by ideal rigid joints, clutches and force elements.

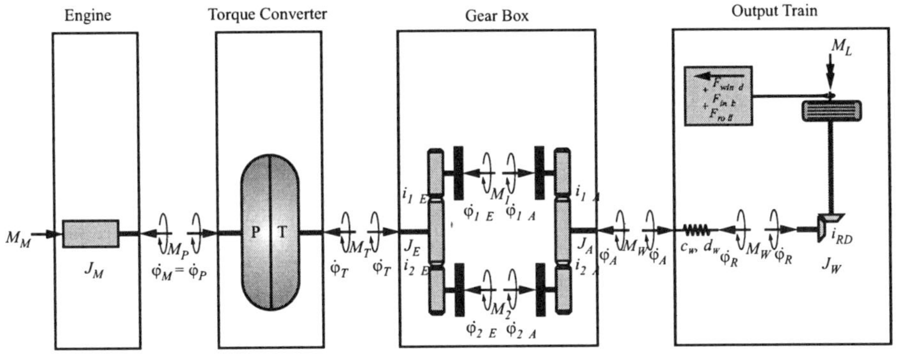

Figure 2: Power train model

For investigations on gear shift comfort the high frequency vibrations of the engine can be neglected. Therefore, the engine can be modeled as a rotating rigid body. The hydrodynamic torque converter consists of a pump (P) which is connected to the input shaft and a turbine (T) which is connected to the output shaft (Figure 2).

The gear box investigated here has five speeds. For simulation of the gear shift process from the first to the second gear we use the reduced model shown in Figure 3. Thereby only gearwheels and shift elements that are in the power flow during the first (index 1) and the second gear (index 2) are considered. J_E and J_A represent the moments of inertia of the driving side (input) and the driven side (output) respectively. The transmission ratio $i_1 = i_{1E} i_{1A}$ of the first gear results from the gear transmission ratios i_{1E} before and i_{1A} after the corresponding shift element. The same applies to the transmission ratio $i_2 = i_{2E} i_{2A}$ of the second gear.

The shift element which is active in the first gear is a one-way clutch. The transition to the second gear is achieved by engaging a wet clutch. Once the clutch pressure is high enough for the wet clutch to transmit the entire input drive torque, the one-way clutch releases and its relative speed becomes positive. The pressure in the wet clutch is increased continuously until its relative speed disappears. This indicates that the clutch sticks and the second gear is fully engaged.

The model of the output train includes the cardan shaft up to the wheels and the chassis of the vehicle. The elasticity of the shafts and tires are modeled as torsional force element with the spring coefficient c_W and the damping factor d_W (Figure 2).

3 MATHEMATICAL DESCRIPTION

Making use of the formula for the angular momentum, the equations of motion of the entire mechanical model (Figure 2) can be formulated as

$$\begin{pmatrix} \ddot{\varphi}_M \\ \ddot{\varphi}_T \\ \ddot{\varphi}_A \\ \ddot{\varphi}_R \end{pmatrix} = \begin{pmatrix} \frac{1}{J_M}(M_M(1-\beta) - M_P) \\ \frac{1}{J_E}\left(M_T + \frac{1}{i_{1_E}} M_1 + \frac{1}{i_{2_E}} M_2\right) \\ \frac{1}{J_A}(-M_W - i_{1_A} M_1 - i_{2_A} M_2) \\ \frac{1}{J_W}(M_W - M_L) \end{pmatrix} . \tag{1}$$

The engine torque is interpolated from a measured two-dimensional characteristic map (Figure 4) as a function of the throttle opening α_{TH} and the engine speed $\dot{\varphi}_M$

$$M_M = M_M(\alpha_{TH}, \dot{\varphi}_M) \tag{2}$$

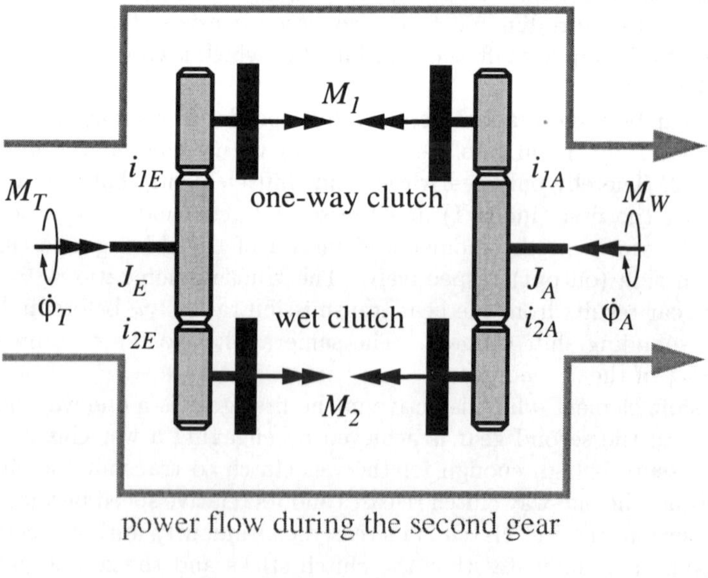

Figure 3: Gear box model

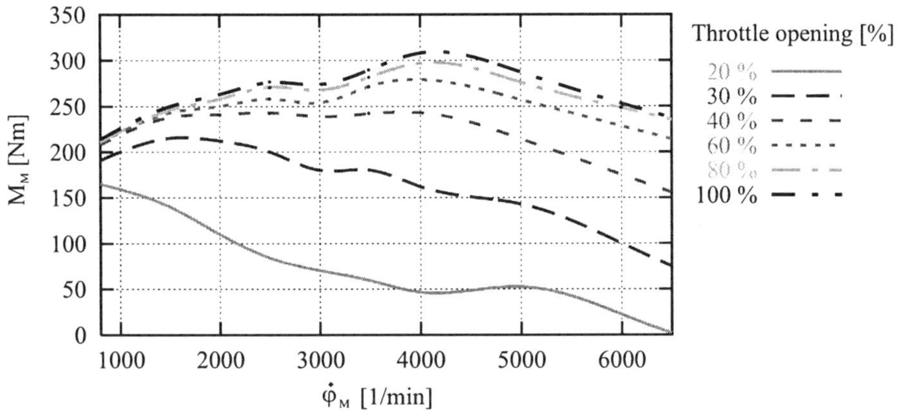

Figure 4: Characteristic map of the engine

The factor β represents the load-reduction applied to the engine during the gear shift process. Outside the shift process β is set to zero.

The pump torque itself can be calculated using a torque characteristic $M_{PC}(\nu)$

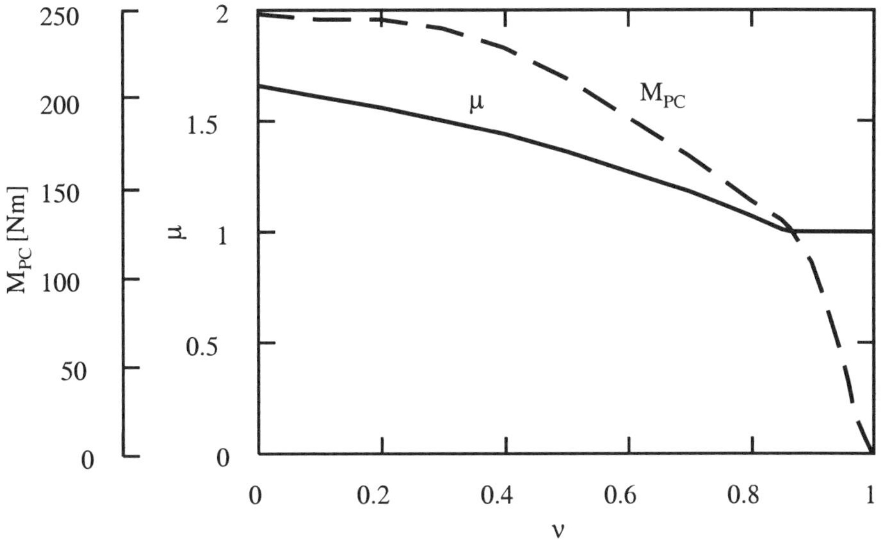

Figure 5: Characteristic map of the hydrodynamic torque converter

measured for a constant pump speed $\dot{\varphi}_{PC} = 2000\ 1/min$

$$M_P = M_{PC}(\nu)\left(\frac{\dot{\varphi}_P}{\dot{\varphi}_{PC}}\right)^2 . \tag{3}$$

With the definition of the velocity ratio ν and torque ratio μ (Figure 5)

$$\nu = \frac{\dot{\varphi}_T}{\dot{\varphi}_P},\ \mu(\nu) = \frac{M_T}{M_P} \tag{4}$$

and for given input and output speeds $\dot{\varphi}_P$ and $\dot{\varphi}_T$ the turbine torque can be calculated as

$$M_T = \mu(\nu) M_P . \tag{5}$$

The determination of the torques M_1 and M_2 of the shift elements depends on their operating state. The relative velocity of the one-way clutch can be defined as

$$\Delta\dot{\varphi}_1 = \dot{\varphi}_A i_{1_A} - \frac{\dot{\varphi}_T}{i_{1_E}} . \tag{6}$$

The one-way clutch allows a relative motion in one direction and blocks it up in the other direction. By defining this free direction as positive the two operating states can be described as follows:

- $\Delta\dot{\varphi}_1 = 0 \Leftrightarrow$ one-way clutch blocks
- $\Delta\dot{\varphi}_1 > 0 \Leftrightarrow$ one-way clutch unblocks

In the first gear we obtain the constraint $\Delta\ddot{\varphi}_1 = 0$ for the one-way clutch. Substituting the second and third equations of (1) in this constraint for its relative acceleration and taking into account that the one-way clutch cannot transmit a negative torque, we get

$$M_1 = \frac{1}{\frac{J_A}{i_{1_E}^2} + i_{1_A}^2 J_E}\left(-\frac{J_A}{i_{1_E}} M_T - i_{1_A} J_E M_W - \left(\frac{J_A}{i_{1_E} i_{2_E}} + i_{1_A} i_{2_A} J_E\right) M_2\right) \geq 0 . \tag{7}$$

As long as $M_1 > 0$ the one-way clutch remains blocked ($\Delta\dot{\varphi}_1$). Once M_1 reaches zero, the one-way clutch starts rotating and it cannot transmit any torque ($M_1 = 0$) as long as $\Delta\dot{\varphi}_1 > 0$. This transmission behaviour is described by the complementary transmission law

$$\Delta\dot{\varphi}_1 \geq 0 ; \qquad M_1 \geq 0 ; \qquad \Delta\dot{\varphi}_1 M_1 = 0 . \tag{8}$$

The transition to the second gear is carried out by engaging the second shift element, which is a wet clutch. Depending on its relative speed

$$\Delta\dot{\varphi}_2 = \dot{\varphi}_A i_{2_A} - \frac{\dot{\varphi}_T}{i_{2_E}} \tag{9}$$

two operating phases are distinguished

- $\Delta\dot{\varphi}_2 \neq 0 \Leftrightarrow$ wet clutch slips
- $\Delta\dot{\varphi}_2 = 0 \Leftrightarrow$ wet clutch sticks

The torque transmitted by the wet clutch during the sliding phase is calculated as

$$M_2 = -\text{sign}(\Delta\dot{\varphi}_2)\, \mu_C\, z\, A_p\, r_m\, p_C \tag{10}$$

with the clutch pressure p_C, the dynamic friction coefficient μ_C, the piston surface A_p, the number of friction surfaces z and the mean friction radius r_m.

Once the control pressure is high enough for the wet clutch to transmit the whole driving torque, the one-way clutch releases. This transition occurs when the torque M_1 becomes zero and the relative acceleration $\Delta\ddot{\varphi}_1$ becomes positive.

The second gear is engaged when the wet clutch sticks. The sticking point is characterized by the condition $\Delta\dot{\varphi}_2 = 0$. During the sticking phase the clutch torque represents a static friction torque and is limited by the inequality

$$-\mu_{C_0}\, A_p\, z\, r_m\, p_C \leq M_2 \leq \mu_{C_0}\, A_p z\, r_m\, p_C \tag{11}$$

where μ_{C_0} is the static friction coefficient of the clutch.

Substituting the second and third equations of (1) into the constraint for sticking $\Delta\ddot{\varphi}_2 = 0$ we obtain

$$M_2 = \frac{1}{\frac{J_A}{i_{2_E}^2} + i_{2_A}^2\, J_E} \left(-\frac{J_A}{i_{2_E}} M_T - i_{2_A} J_E M_W \right) . \tag{12}$$

One should note that, as mentioned above, the one-way clutch releases before the wet clutch sticks. Therefore M_1 is set to zero while evaluating equation (12).

The torque M_W of cardan shaft can be calculated as

$$M_W = c_W(\varphi_A - \varphi_R) + d_W(\dot{\varphi}_A - \dot{\varphi}_R) . \tag{13}$$

The load torque M_L results from the wind force, rolling force and the inclination force

$$M_L = (F_{wind} + F_{roll} + F_{incl})\, \frac{r_R}{i_{RD}} \tag{14}$$

where

$$F_{wind} = \frac{1}{2}\, \rho\, c_w\, A \left(\dot{\varphi}_R \frac{r_R}{i_{RD}} \right)^2 \tag{15}$$

$$F_{roll} = \mu_s\, m\, g\, \cos\gamma \tag{16}$$

$$F_{incl} = m\, g\, \sin\gamma \tag{17}$$

with the wheel radius r_R, the transmission ratio of the rear differential i_{RD}, the air density ρ, the air resistance coefficient c_w, the vehicle front surface A, the mass of the vehicle m, the gravitation g, the road inclination γ and the rolling coefficient μ_s.

4 MODEL VERIFICATION

The comparison between simulations performed with the presented model and measurements carried out during a gear shift process from the first to the second gear shows a very good agreement (Figure 6). Therefore, the mechanical model of the power train can be used to develop a model-based optimization of the gear shift process in automatic transmissions. In the simulation the control data used in the car were implemented and used as reference in the following sections.

5 OPTIMIZATION PROBLEM

5.1 Optimization parameters

The clutch pressure as well as the engine load-reduction are determined in the electronic control unit by interpolation from different meshed characteristic maps. The parameters which must be optimized are the control pressures p_i at the sampling points of the characteristic map for the pressure p_C, which depends on the turbine torque and angular velocity, and the load-reduction factor β_i at the sampling points of the characteristic map for β, which depends on the engine torque and angular velocity. Therefore, the optimization parameter vector can be defined as follows

$$\bar{\boldsymbol{p}} = \left(p_1, \ldots, p_{n_p}, \beta_1, \ldots, \beta_{n_\beta}\right)^T \ . \tag{18}$$

The goal of the optimization is to find a parameter vector $\bar{\boldsymbol{p}}$, such that a system governed by equation (1) shows an optimal performance with respect to the optimization criteria.

5.2 Optimization criteria

The evaluation of the gear shift process with respect to the passengers comfort is a subjective issue and varies from driver to driver. Nevertheless investigations in [5] have shown that there is a correlation between the comfort and the jerk $\dot{a}$ induced by the acceleration change. While it is not easy to find a formula to quantify and specify a comfort mark, i.e. a scale between 1 (very poor) and 10 (excellent), which matches the subjective feeling of the driver, it can generally

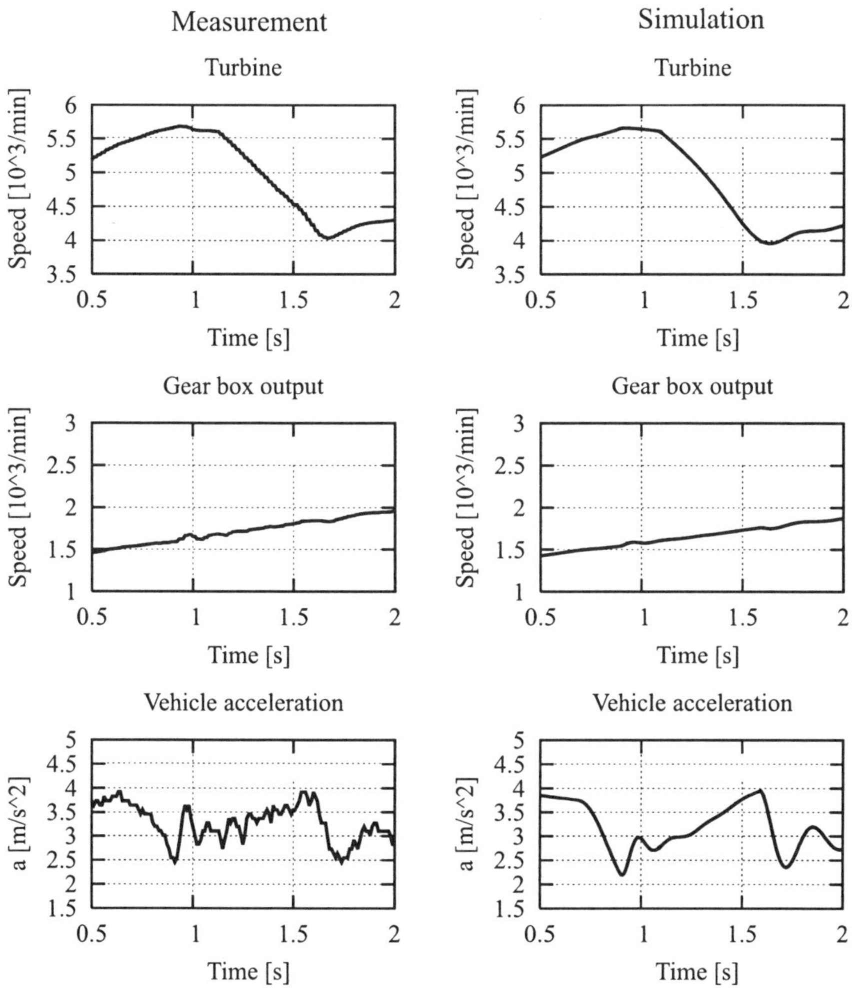

Figure 6: Comparison of simulation and measurement

be stated that the smoother the acceleration change is and the smaller the peaks of the jerk are, the more comfortable the gear shifting is.

Vehicle acceleration

A possible optimization criterion for evaluating the comfort can be formulated by minimizing the difference between the actual acceleration a and a desired acceleration profile $a_{desired}$. The definition of the desired acceleration profile is a very important issue for the quality of the optimization. It must be an optimal and physically reasonable profile. An example for a desired acceleration profile is given in figure 7.

Once the desired profile is defined the cost criterion for the acceleration can be formulated as

$$G_1 = \int_{t_b}^{t_e} (a - a_{desired})^2 \, dt \rightarrow \min \ . \tag{19}$$

where t_b and t_e characterize the beginning and the end of the evaluation time interval.

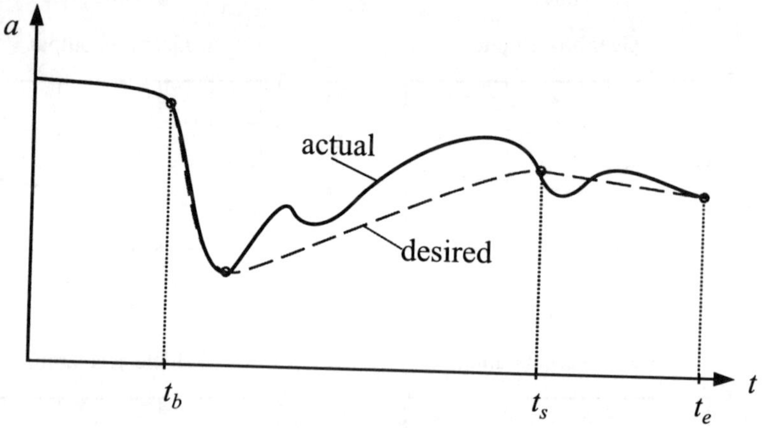

Figure 7: Example for the definition of a desired acceleration profile

Vehicle jerk

Another possibility to evaluate the comfort is a cost function for minimizing the jerk

$$G_2 = \int_{t_b}^{t_e} \dot{a}^2 dt \rightarrow \min \ . \tag{20}$$

Instead of deriving a differential equation for the jerk $\dot{a}$, the acceleration a is used, which is available in discrete form after the numerical integration of the equations of motion (1). The jerk can be calculated using the Euler-step method

$$\dot{a}_i \approx \frac{\Delta a_i}{\Delta t_i} = \frac{a_{i+1} - a_i}{\Delta t} \ . \tag{21}$$

The criterion G_1 is then formulated as follows:

$$G_2 = \sum_1^n \dot{a}_i^2 \Delta t = \sum_1^n \frac{(\Delta a_i)^2}{\Delta t} \ . \tag{22}$$

Gear shift duration
The minimization of the comfort criteria, especially G_2, would lead to a very long gear shifting duration because the logical result of this optimization problem is to engage the clutch of the target gear with a very smooth and slow rising pressure in order to get a very slow transition in the dynamical behaviour of the power train. As a result we get a smooth acceleration but a very long gear shift process. This increases the thermal load in the clutch and negatively affects the dynamic response of the vehicle especially while overtaking. Therefore, a third objective function which evaluates the duration of the gear shift process is introduced

$$G_3 = t_s - t_b \Longrightarrow \min \ . \tag{23}$$

5.3 Optimization constraints

In order to obtain results which can be applied to the real system, certain constraints have to be imposed on the optimization problem. The equations of motion of the system represent constraints for the dynamic behaviour and response upon the input from the control unit in the form of equality equations. The hydraulic system can provide the wet clutches with a maximum pressure of p_{max}. Therefore all pressure values at the sampling points of the characteristic map for the control pressure must be limited to

$$0 \leq p_i \leq p_{max} \ . \tag{24}$$

This ensures that the optimized pressure can be provided by the hydraulic system.

Furthermore the engine electronics allows a maximum load-reduction of β_{max} during the gear shift process. This restriction is taken into account by the limitation of the parameters of the characteristic map of the load-reduction factor

$$0 \leq \beta_i \leq \beta_{max} \ . \tag{25}$$

In order to avoid overdriving situations for the engine especially during a gear shift operation under full load, the engine angular velocity must be limited to a maximum angular velocity $\dot{\varphi}_{max}$

$$\dot{\varphi}_M \leq \dot{\varphi}_{max} \ . \tag{26}$$

Problem formulation

The characteristic maps for the pressure and for the engine load-reduction should be optimal for any possible load case. Therefore, we need to consider as many load cases as possible for the optimization. This is achieved in this work by taking into account n_{LC} simulations with different throttle openings and different shift programs. Since we can evaluate three criteria for each load case, we get a $3n_{LC}$-vector of the normalized criteria for the entire problem

$$\boldsymbol{G} = \left(\frac{G_{11}}{G_{11,0}}, \frac{G_{21}}{G_{21,0}}, \frac{G_{31}}{G_{31,0}}, \ldots, \frac{G_{1j}}{G_{1j,0}}, \frac{G_{2j}}{G_{2j,0}}, \frac{G_{3j}}{G_{3j,0}}, \ldots, \right. \tag{27}$$

$$\left. \frac{G_{1n_{LC}}}{G_{1n_{LC},0}}, \frac{G_{2n_{LC}}}{G_{2n_{LC},0}}, \frac{G_{3n_{LC}}}{G_{3n_{LC},0}}\right)^T \tag{28}$$

The complete vector problem for the gear shift optimization is

$$\min_{\bar{\boldsymbol{p}}}\{\boldsymbol{G}:\ \boldsymbol{g}_1 = \boldsymbol{0};\ \boldsymbol{g}_2 \leq \boldsymbol{0}\}\ . \tag{29}$$

The equality equations $\boldsymbol{g}_1$ represent the equations of motion describing the dynamics of the drive train (eq. 1). The inequality for $\boldsymbol{g}_2$ includes all restrictions of the optimization parameters (eq. 24 and 25) and the engine angular velocity (26).

The vector problem can be reduced to a scalar problem using the method of the objective weighting [1]. This is performed in a first step by introducing the weighting factors $\boldsymbol{w}_j = (w_{1j}, w_{2j}, w_{3j})^T$ for each load case such that

$$0 \leq w_{ij} \leq 1; \qquad \sum_{i=1}^{3} w_{ij} = 1 \tag{30}$$

and defining the scalar preference function for each load case

$$G_j = w_{1j}\frac{G_{1j}}{G_{1j,0}} + w_{2j}\frac{G_{2j}}{G_{2j,0}} + w_{3j}\frac{G_{3j}}{G_{3j,0}} \tag{31}$$

In a second step the criteria for the different load cases are weighted and added together in a scalar preference function for all load cases in the form

$$G = \sum_{j=1}^{n_{LC}} W_j G_j \tag{32}$$

with the weighting factors

$$0 \leq W_j \leq 1; \qquad \sum_{j=1}^{n_{LC}} W_j = 1 \tag{33}$$

The solution of the optimization can be obtained by solving the scalar substitute problem

$$\min_{\bar{\boldsymbol{p}}}\{G:\ \boldsymbol{g}_1=\mathbf{0};\ \boldsymbol{g}_2\leq\mathbf{0}\} \tag{34}$$

for each set of weighting factors $\{w_{ij}\}$ and $\{W_j\}$, which fulfills (30) and (33).

This nonlinear optimization problem is solved by a sequential quadratic programming algorithm. By calculating the derivatives of the objective functions analytically instead of using a numerical method, a significant improvement is achieved in terms of convergence.

6 RESULTS

Optimization results have shown that the passengers comfort and the duration of the gear shift operation represent contrary issues. This means that an improvement in one criterion leads to a deterioration in the other one. Therefore the optimization problem represents be leads to a pareto-optimal solution region (figure 8). A reasonable solution can be found by defining a compromise between the two issues. The compromise used in this work consists of improving the comfort without increasing the reference gear shift duration.

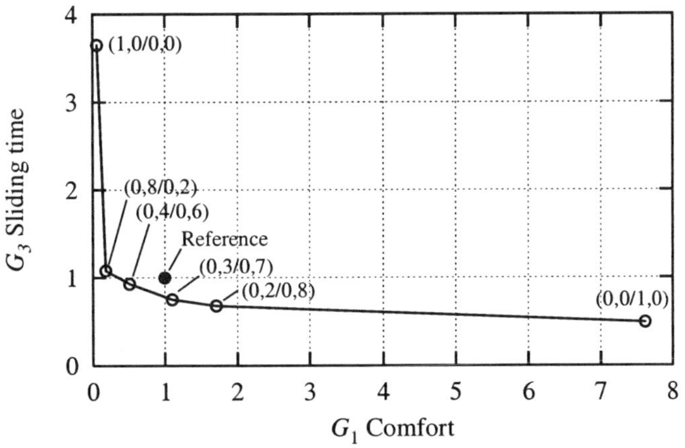

Figure 8: Pareto optimal region for the gear shift in the criteria space (G_1, G_3)

First, we consider optimization results achieved using the criterion for the acceleration G_1 and the criteria for the jerk G_2. Both results were achieved for the full load case (100% throttle opening) in combination with the criterion G_3 for the gear shift duration. Figure 9 shows that both comfort criteria lead to a remarkable improvement of the smoothness of the acceleration and the jerk.

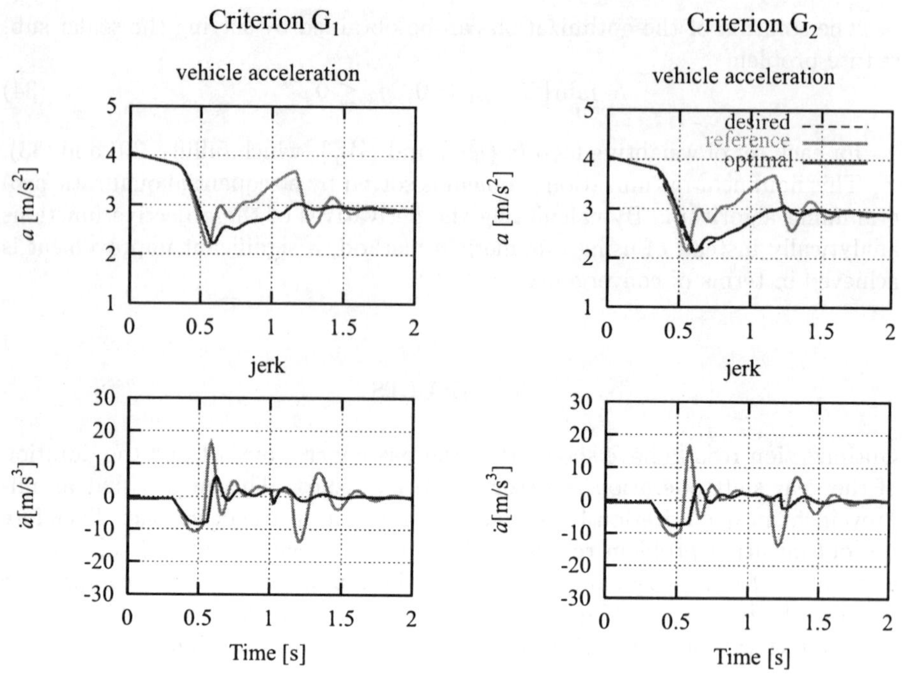

Figure 9: Comparison of optimization results using the criteria (G_1, G_3) and (G_2, G_3), 100% load

It can also be seem that the gear shift duration decreases slightly. The pareto-optimal solution was obtained in both cases by using the weighting factors $w_1 = w_2 = 0.7$ for the criteria G_1 and G_2 and $w_3 = 0.3$ for G_3.

Figure 10 shows the optimized vehicle acceleration and jerk compared to the reference for 30%, 60% and full load using G_2 and G_3. It can be seen that significant improvements in terms of passengers comfort and gear shift duration compared to the reference case are achieved. The plot for the relative velocity of the wet clutch in figure 10 shows that the gear shift duration was almost constant compared to the reference.

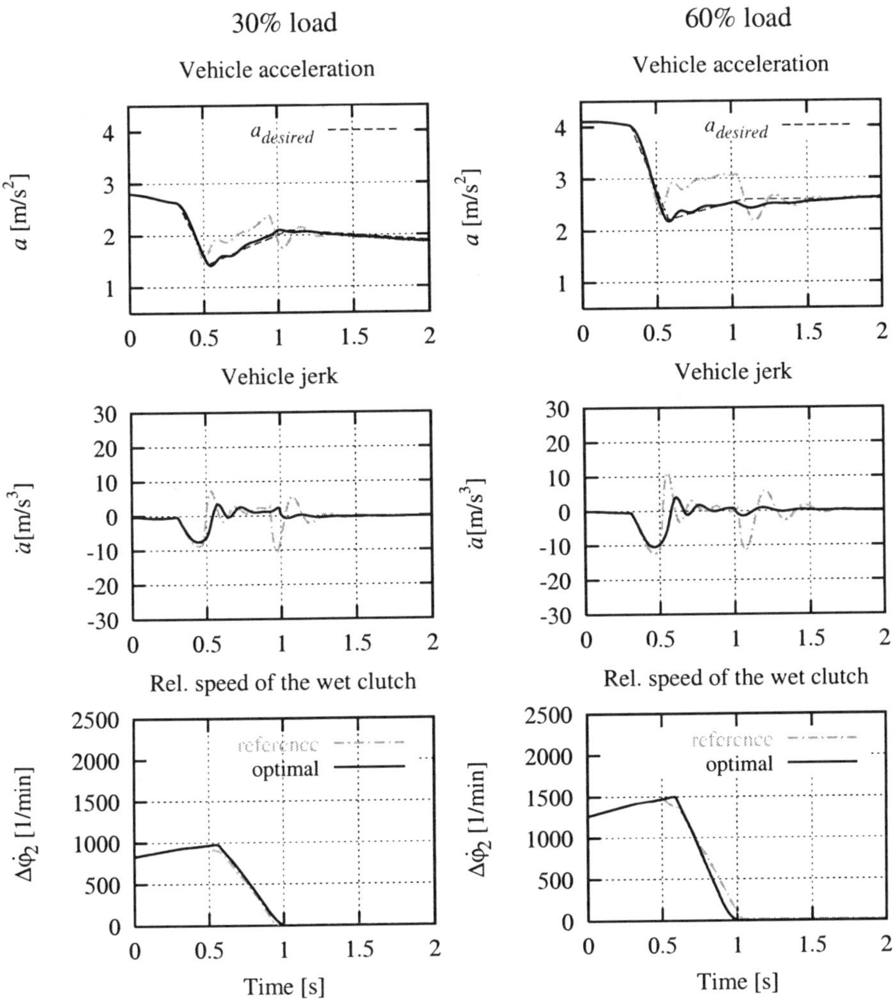

Figure 10: Optimization results for 30% and 60% load

7 CONCLUSIONS

In this paper a model based method has been presented for the optimization of the control data for the clutch pressure as well as those for the engine load-reduction in order to improve the performance of a power train with an automatic transmission. The passengers comfort and the gear shift duration were considered as optimization criteria. It turned out that both objectives are contrary. Therefore the optimization leads to a pareto-optimal solution. By choosing a set of weighting factors, which emphasizes the importance of the comfort a significant improvement in terms of this criterion was achieved for different load cases.

REFERENCES

1. Eschenauer, H. A., Koski, J., Osyczka, A. 'Multicriteria Optimization - Fundamentals and Motivation', *Multicriteria Design Optimization*, Springer, Berlin, pp. 1-32.
2. Förster, H.-J. 'Getriebeschaltung ohne Zugkraftunterbrechung', In: *Automobil Industrie*, pp. 60–76, 1962.
3. Gebert, J. 'Adaptive Parametervariation bei Getriebesteuerungen zur Optimierung des Schaltkomforts'; *VDI Fortschritt-Berichte*, Reihe 12, Nr. 424, Düsseldorf: VDI Verlag.
4. Haj-Fraj, A., Pfeiffer, F. 'Dynamics of Gear Shift Operations in Automatic Transmissions', *Proceedings of the First International Conference on the Integration of Dynamics, Monitoring and Control for the 21st Century.* Manchester, UK, 1–3 September, pp. 29–35.
5. Küçükay, F., Renoth,F. 'Intelligente Steuerung von Automatikgetrieben durch den Einsatz der Elektronik', *ATZ Automobiltechnische Zeitschrift* 96, No. 4, pp. 228–235.
6. Winchell, F.J., Route, W.D. 'Ratio Changing the Passenger Car Automatic Transmission', *Design Practices–Passenger Car Automatic Transmissions*, Part 1, SAE AE-1, 57–80, 1962.